W0232096

СЕЙСМИЧЕСКИЙ ЭФФЕКТ
ПОДЗЕМНЫХ ВЗРЫВОВ

SEISMICHESKII EFFEKT PODZEMNYKH VZRYVOV
SEISMIC EFFECTS OF UNDERGROUND EXPLOSIONS

**Transactions (Trudy) of the
O. Yu. Shmidt Institute of Geophysics No. 15 (182)**

SEISMIC EFFECTS OF
UNDERGROUND EXPLOSIONS

Authorized translation from the Russian

SPRINGER SCIENCE+BUSINESS MEDIA, LLC
1962

ISBN 978-1-4899-5060-4   ISBN 978-1-4899-5058-1 (eBook)
DOI 10.1007/978-1-4899-5058-1

Сейсмический эффект подземных взрывов
Труды Института физики земли
им. О. Ю. Шмидта

Library of Congress Catalog Card Number: 61-17729

Copyright 1962 Springer Science+Business Media New York
Originally published by Consultants Bureau Enterprises, Inc. in 1962
Softcover reprint of the hardcover 1st edition 1962

No part of this publication may be reproduced
in any form without written permission
from the publisher

CONTENTS

RESULTS OF SEISMIC OBSERVATIONS
ON UNDERGROUND NUCLEAR AND TNT EXPLOSIONS

I. P. Pasechnik, S. D. Kogan, D. D. Sultanov, and V. I. Tsibul'skii

This paper describes the nature of the seismic records obtained in the USA and the USSR from underground explosions produced in the USA under the designations Rainier, Tamalpais, Logan, and Blanca at the test site in Nevada in 1957 and 1958. Data are furnished on the transit times of the seismic waves, on the relationship of amplitude of the longitudinal waves to the epicentral distance, and also on the numerical magnitudes. It is shown that the spatial distribution of positive and negative signs (directions) of the first movement on records of stations surrounding the epicenter differs substantially from the analogous distribution for the overwhelming majority of earthquakes. A method is proposed for correlating the initial parts of the record; this aids in distinguishing the first movement. Values are given for the periods of body and surface waves from the explosions. The difference in periods of surface waves from explosions and earthquakes, comparable in energy level, must be one of the criteria for recognizing explosions. Evaluations are given for the energy of seismic waves arising during underground nuclear and TNT explosions.

Introduction

In order to solve the problem of recognizing underground nuclear explosions much interest has been aroused in making thorough studies of seismic records of such explosions from seismic stations in the USA, the USSR, and other countries.

Until the present there have been published a number of papers with descriptions and analyses of seismic data obtained only in the USA [14-16, 32, 34].

The present paper furnishes information on seismic observations in the USSR and in other countries and analyzes this information from the viewpoint of possible detection and recognition of underground nuclear explosions. In connection with this, studies are made of seismic data obtained in the USSR from underground TNT explosions.

The paper makes use of and reworks the seismic material communicated by the USA and USSR delegations at Geneva in 1959 at the Conference for stopping tests of nuclear weapons; sources in the literature are also used.

1. Actual Data on Underground Nuclear Explosions Produced in the USA in 1957 and 1958

At present we have at our disposal information on six underground nuclear explosions produced in the USA at the site in Nevada. The first underground explosion, called the Rainier explosion, of the Plumbbob series of tests, was set off in 1957; the other five explosions were part of the Hardtack II tests on nuclear devices in 1958. Information on the coordinates of the epicenters, on the depth, time, and strength (or yield) of the explosions, from published data [14-20, 28-32, 34, 35], is presented in Table 1.

TABLE 1. Data on Underground Nuclear Explosions

Name of explosion	Date	Time GCT h m s	North latitude	West longitude	Depth, m	Approx. yield, kt
Evans	Oct. 29, 1958	00 00 00.1	37°11′42″	116°12′17″	280	0 055
Tamalpais	Oct. 8, 1958	22 00 00.1	37 11 43	116 12 01	110	0.072
Neptune	Oct. 14, 1958	18 00 00.1	37 11 38	116 11 59	33	0.090
Rainier	Sept. 19, 1958	16 59 59.4	37 11 45	116 12 11	283	1.7
Logan	Oct. 16, 1958	06 00 00.1	37 11 03	116 12 04	278	5
Blanca	Oct. 30, 1958	15 00 00.1	37 11 09	116 12 07	280	19

All six nuclear explosions were detonated in the same geological environment, in a layer of tuff about 600 m thick overlying dense crystalline limestone.

Various papers indicate different values of TNT equivalents for the underground nuclear explosions. In the remainder of this paper the TNT equivalents indicated in Table 1 will be used for the explosions.

2. The Network of Observations for the 1958 Explosions

According to the paper [34] seismic observations of the Hardtack II series of underground explosions were made at special temporary stations in a number of places distributed along a line eastward from the test site in Nevada to Arkansas and thence northeastward to the state of Maine. The nearest station was approximately 100 km from the epicenter of the explosion, the most distant somewhat over 4000 km away. The station sites were selected by geologists; the stations were located on outcrops of crystalline rocks remote from sources of man-made noise. The positions of the stations are shown in Fig. 1, which is reproduced from the paper [34].

Eight of the temporary stations were kept at fixed sites for all the series of tests in order to shed light on the relationship between amplitude and yield of an explosion. Ten stations were kept mobile and were shifted between shots in order to increase the number of observations and shed more light on the relationship between amplitude and distance. After the Logan shot most of the mobile stations were shifted about 100 km nearer the source. The largest shots were also recorded at many of the regularly operating seismograph stations in the USA and in other countries.

3. Apparatus

The underground nuclear explosions of the Hardtack II series were recorded at specially established stations by Benioff short-period vertical seismographs. According to the paper [34], these seismographs were designed with a free period of one second and a damping constant (D_1) of about 0.7. The seismometers were couples with galvanometers having a free period of 0.2 sec and a damping constant (D_2) of 1.0. The coupling coefficient of the pickup link was $\sigma^2 \approx 0$. The frequency characteristics of the Benioff vertical seismograph for displacements are shown in the paper [14] and are reproduced here in Fig. 2 (curve 1).

It may be seen from an examination of the frequency characteristics that the magnification of these instruments is irregular. The greatest magnification is at a period of 0.3 sec; at periods of 0.1 and 0.8 sec the magnification declines to but half the maximum, and at periods of 1 sec, it is but one-third. After that the magnification decreases sharply toward longer-period oscillations. Thus, the apparatus employed had a comparatively narrow range, with a passband of 0.13-0.7 sec, and, because of this, there was considerable distortion in the shapes of the recorded impulses and, of primary significance, a decrease in the amplitudes of the first arrivals on the records. The pickup circuit recorded not the displacement of the ground but some function approximating the velocity of the displacement.

The magnification of seismographs for periods of one second range from 1000 to 272,000 for different distances from the source. Data on magnification of the apparatus are shown on copies of the seismograms and are tabulated in Table 2. The magnification at each station is known with a precision of approximately 5-10% [34], since each

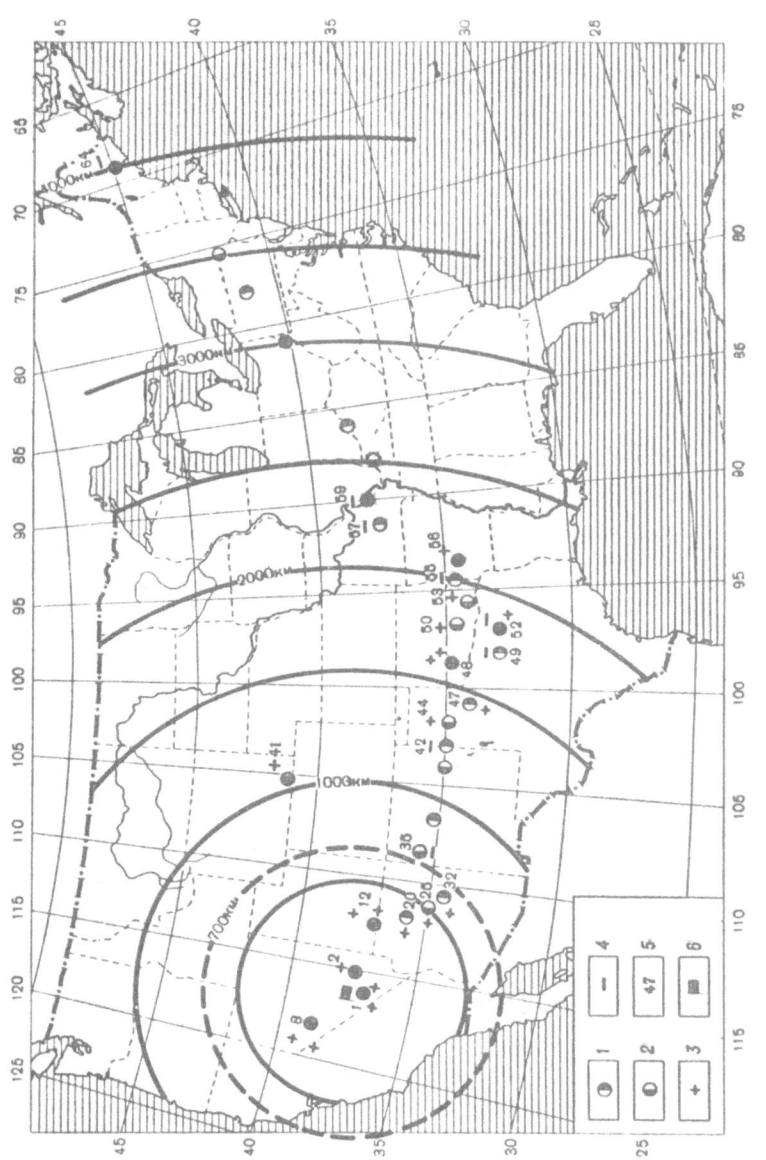

Fig. 1. Distribution of the special stations for the Blanca and Logan shots [34]. 1) Blanca; 2) Logan; 3) compressional wave; 4) rarefactional wave; 5) figures correspond to sequential numbers in list of Table 5; 6) explosion site.

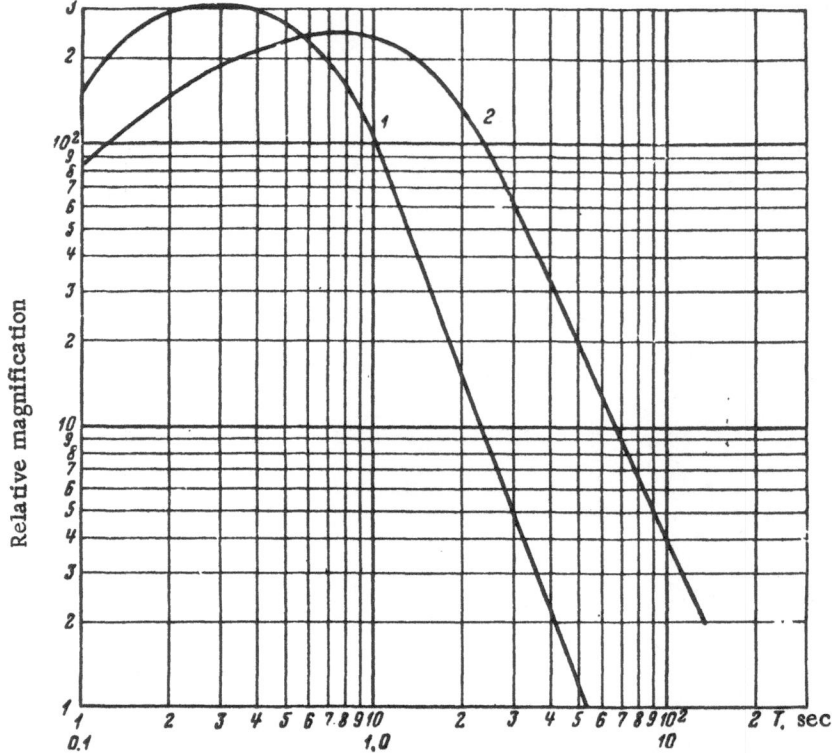

Fig. 2. Frequency characteristics of seismographs: 1) Benioff; 2) SVK-M.

special station had a device for calibration. There were also devices at the stations for determining the direction of displacement of the light beam under operating conditions and means of recording time signals on a continuous basis from a radio station.

Different apparatus was employed at the regularly operating seismograph stations in the USA. At these stations most of the seismic records of the explosions were obtained on Benioff and Sprengnether seismographs and on Wood-Anderson torsion seismographs. Unfortunately, we did not know all the parameters for the designs and magnifications of the instruments.

The seismic records of the Blanca and Logan shots were obtained at Soviet seismic stations by SVK-M vertical seismographs, which have a broader passband, 0.2-2.0 sec, than the Benioff seismographs. A description of SVK-M and SGK-M seismographs is given in the papers [8] and [13].

The SVK-M seismographs were adjusted to a free period of 2.5 sec, and, at a station 10,080 km from the source, a period of 4 sec; they were coupled with galvanometers with periods of 1.1 sec. The damping factor of the seismographs (D_1) was about 1.5 and of the galvanometers (D_2), about 3. The coupling coefficient (σ^2) did not exceed 0.3.

The frequency characteristic for displacement of the SVK-M seismograph, determined at one of the Soviet seismic stations, is shown in Fig. 2 (curve 2). As seen from an examination of this curve, the greatest magnification of the seismograph lay at periods near 1 sec; at periods of 0.2 and 1.2 sec the magnification of the instrument was less by a factor of approximately 1.5; and at periods greater than 1.2 sec the magnification decreased rapidly. The magnification at the maximum, at periods near 1 sec, was 20,000-30,000.

4. Characteristics of the Seismograms

At present the most careful treatment has been afforded seismograms of shocks of the Hardtack II series obtained at stations of the special network on Benioff vertical seismographs. The upward deflection of the light beam on these seismograms corresponds to upward movement of the ground, i.e., to a compressional wave. The records at the special stations were made at a rate of 120 mm/min; a time mark was made every 10 sec, except at minute intervals, when the mark was omitted. Greenwich time is indicated on each seismogram, at the initial point of measurement. The coordinates of the points where the records were obtained are not shown.

4

TABLE 2. Phases Distinguished on Records of Underground Nuclear Explosions of the Hardtack II Series and Published in Seismological Journals (V is the magnification for an oscillating period of 1 sec; the time is given for instant of arrival)

No.	Station	Δ, km	Blanca V	Blanca h m s	Logan V	Logan h m s	Tamalpais V	Tamalpais h m s
1		96.2			1600	$+iP*$ 06 00 16.0 iP_n 17.5 e 29.6	21 200	$+eP*$ 22 00 16.6 P_n 17.7 e 25.7 e 27.5
2		203.4	1000	$+P$ 15 00 31.1	8000	$+iP_n$ 31.5	62 700	$+eP_n$ 32.7 $eP*$ 34.7 e 45.2 e 51.5
3		203.5						eS_n 01 00.0
4		300.6	2500	$+iP_n$ 44.6 $iP*$ 49.9 e 50.9 e 54.2 i 56.6 e 01 04.0 iS_n 20.2 $iS*$ 28.2	12 000	$+iP_n$ 44.7 $iP*$ 50.0 i 51.0 e 01 02.8 eS_n 20.3	103 000	$+eP_n$ 00 44.8 $iP*$ 51.4 i 56.8 e 01 02.1 e 11.5 i 18.5
5		395.1	10 600	$+iP_n$ 00 57.3 $iP*$ 01 06.2 i 24.0 iS_n 42.2 e 50.6 $eS*$ 54.1				

TABLE 2 (Cont'd)

No.	Station	Δ, km	Blanca V	Blanca	h m s	Logan V	Logan	h m s
6		498.9				115 000	$+iP_n$	06 00 10.3
							$eP*$	20.5
							eS_n	02 05.1
							$eS*$	22.9
7		600.0	26 000	$+iP_n$	15 00 23.1	98 800	$+eP_n$	01 37.5
8		714.5					i	45.6
							$iP*$	56.1
							e	02 02.4
							e	14.0
							e	59.1
9		909.0	77 000	P_n	02 01.2			
				e	16.6			
				e	21.0			
10		1036.0	192 000	$+P_n$	00 17.5	272 000	$-eP_n$	00 17.8
				i	31.7		e	27.2
				$iP*$	46.5		i	32.0
				i	03 00.0		$iP*$	46.2
							i	59.8
							eP_n	02 30.9
							$eP*$	03 01.0
							i	08.8
							i	20.5
11		1111.5				114 000		
12		1215.0	98 500	$-iP$	02 44.6			
				i	03 15.3			
				i	19.1			

6

TABLE 2 (Cont'd)

No.	Station	Δ,km	Blanca v	Blanca	h m s	Logan v	Logan	h m s
13		1313.1		e	15 03 55.6	40 000	eP	06 02 53.3
14		1610.0	161 000	—iP i	29.4 04 31.7			
15		1610.1				116 000	+P i i	03 30.4 41.8 04 32.2
16		1707.0	113 500	—iP i	03 39.4 04 47.7			
17		1803.7				175 000	+iP	03 51.0
18		1902.1				172 000	+iP	04 02.9
19		2011	147 000	—iP ι e e	14.3 16.9 33.3 36.5			
20		2111	93 000	—iP i	24.9 50.0			
21		2111.3		—iF	34.3	88 000	P	23.9
22		2208	123 000		39.7			
23		2305*						
24		2506**						
25		2665.0***	44 000	P	05 18.7	67 600	—iP	46.1
26		3017	62 000	—iP	54.5	30 000		

TABLE 2 (Cont'd)

No.	Station	Δ, km	Blanca v	Blanca h m s	Logan v	Logan h m s
27		3017.4			51 900	Shock not recorded
28		3309	92 000	Shock not recorded		
29		3502			105 000	The same
30		4020.6			96 000 ****	»
31		4021	67 000	−iP 15 07 04.9 ****		
32	Tiksi	6890	24 000	−iP 10 25		−P 06 10 23.4
33	Temporary station	8300	23 000	+P 11 44.7		eP 11 43.7
34	Kew	8350		iP 45.6		
35	Uppsala	8410		iP 49		
36	Matsushiro	8800		+iP 12 08		
37	Stuttgart	9020		eP 20		
38	Pruhonice	9180	6000	+P 26.5		
39	Temporary station	10 080	28 000	+P 13 07		
40	Banger Oasis	16 040	9 000	ePKF 19 33		
41	Mirnyi	16 230	12 000	+ePKP 40.3		

*The direction of motion on the photographic record, in contrast to the records of the other stations, is from right to left.

**Noise

***Heavy background of microseisms.

****Heavy background of microseisms.

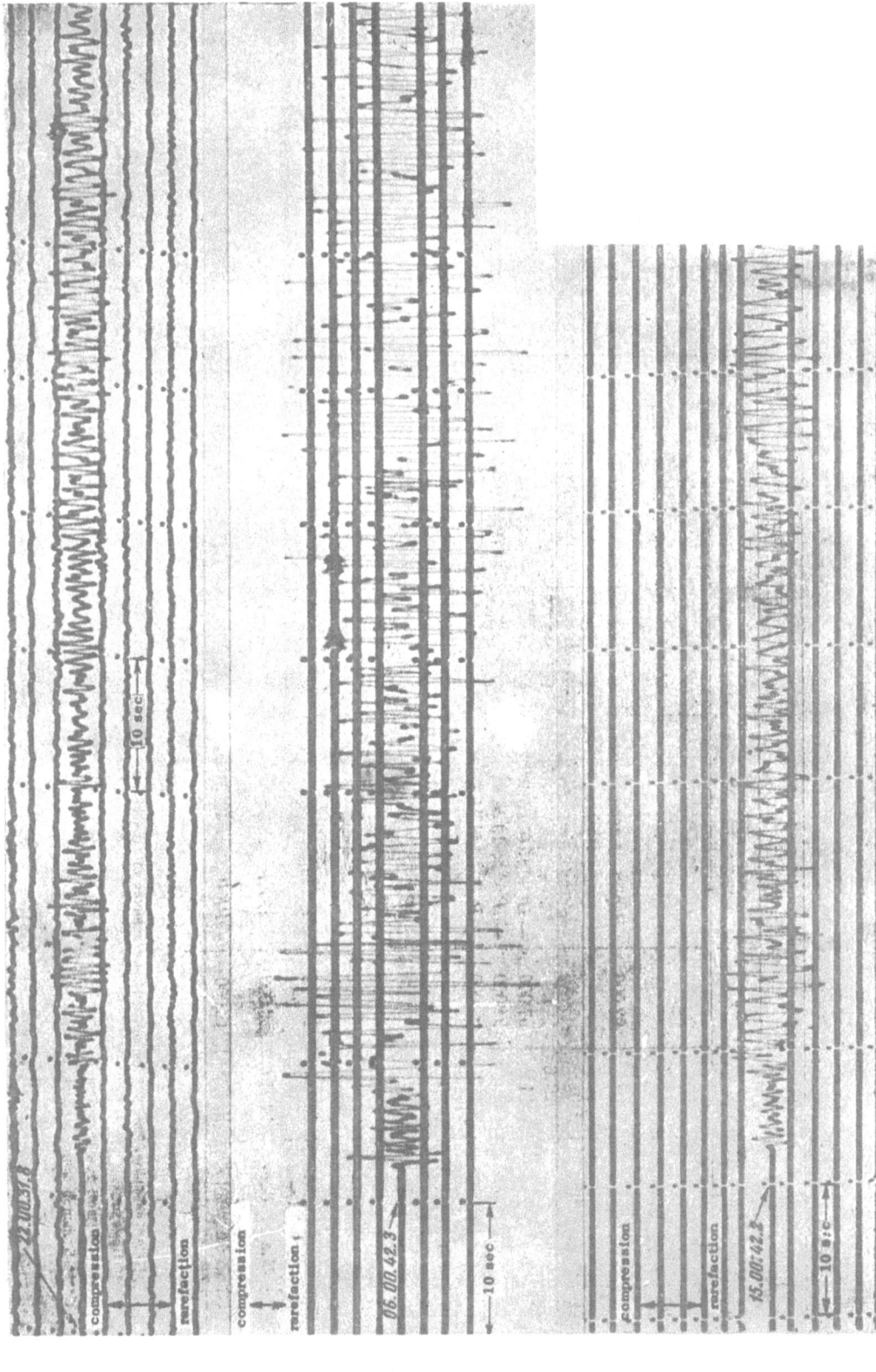

Fig. 3. Seismograms of underground nuclear explosions at a distance Δ = 300 km. a) Tamalpais, V = 103,000; b) Logan, V = 12,000; c) Blanca, V = 2,500.

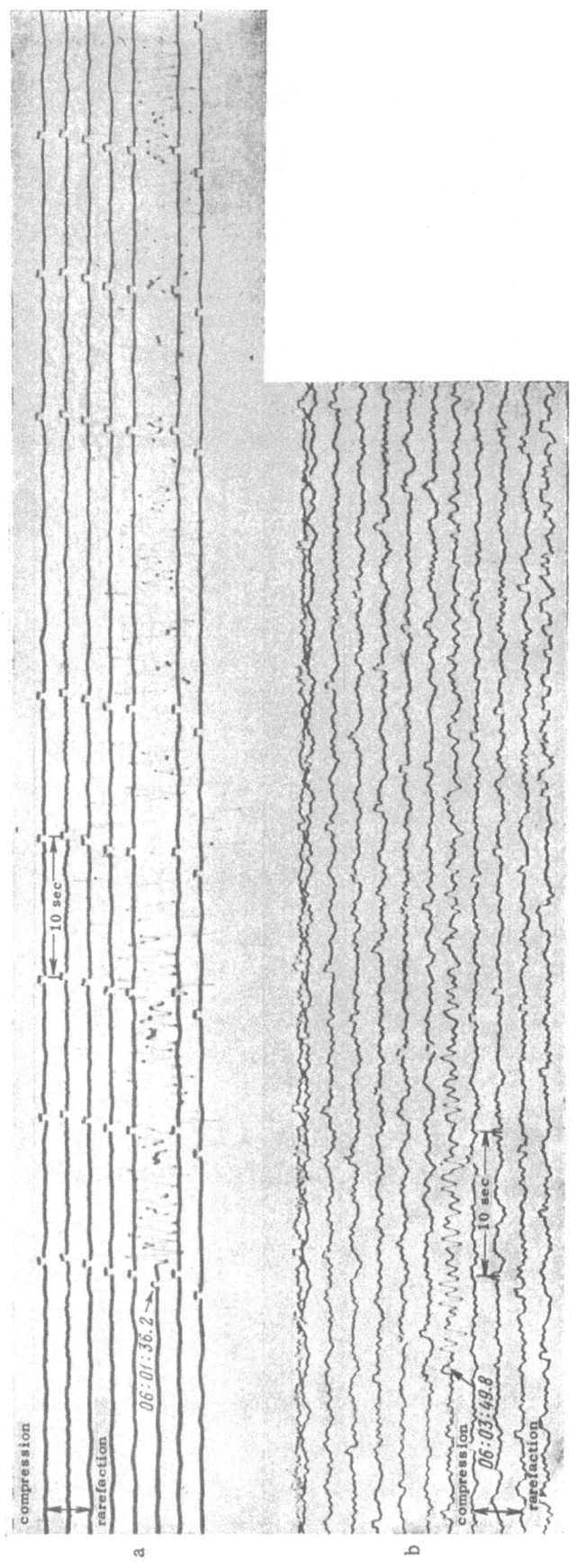

Fig. 4. Seismograms of the Logan shot. **a)** Δ = 714 km, V = 98,800; **b)** Δ = 1804 km, V = 175,000.

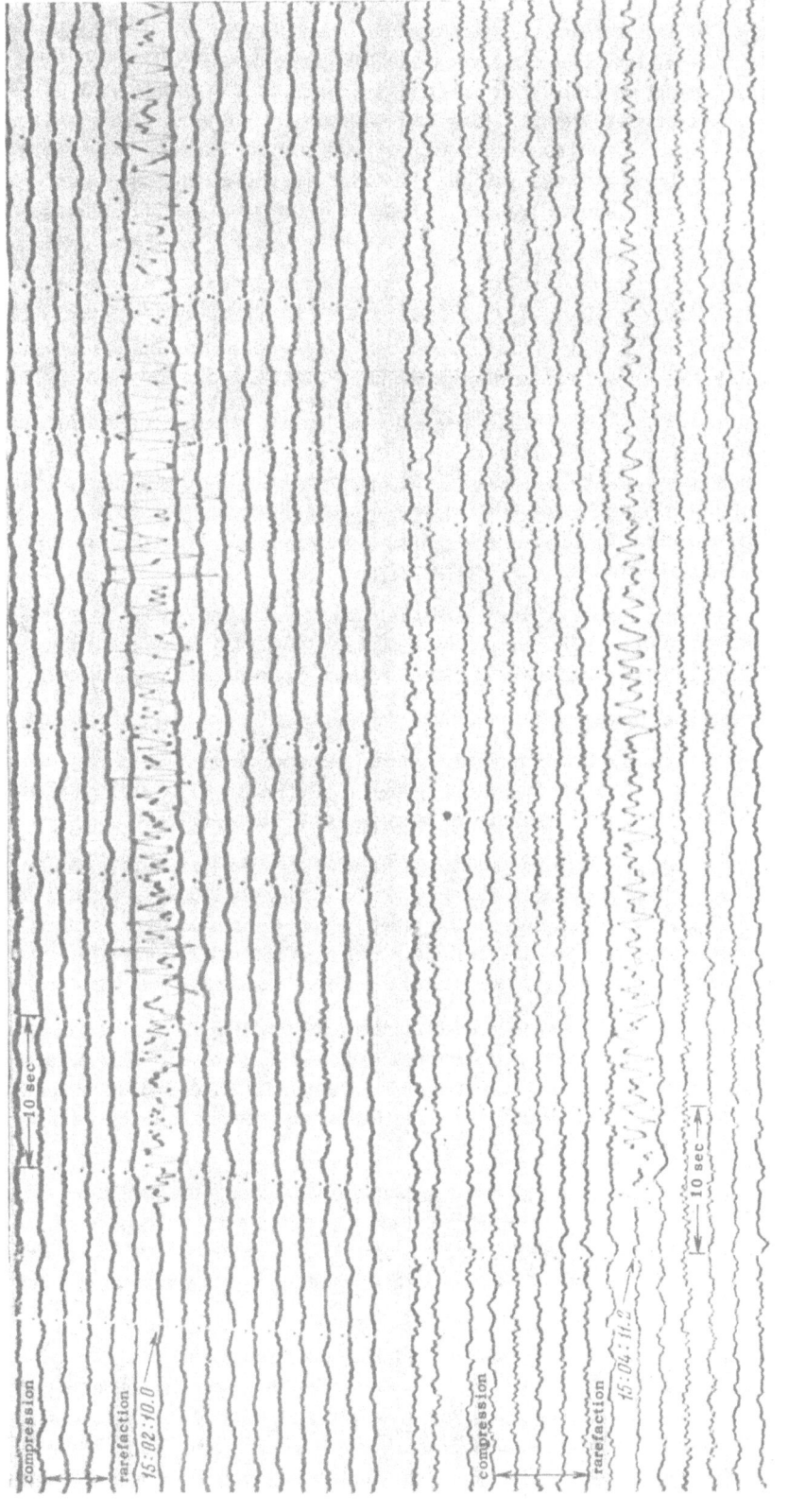

Fig. 5. Seismograms of the Blanca shot, a) Δ = 1036 km, V = 192,000; b) Δ = 2011 km, V = 147,000.

On most of the instruments at the regularly operating seismograph stations in the USA, the rate of recording did not exceed 60 mm/min.

Examples of seismograms of the different explosions are shown in Figs. 3-5. The background of microseisms on the seismograms obtained at stations approximately up to 700 km from the explosion site is insignificant; the records are not complicated by superposed microseisms and the appearance of the Pn wave is clear. At distances beyond 700 km, where the apparatus was adjusted for greater magnification than at the shorter distances, the seismograms are complicated by short-period microseisms with periods chiefly of 0.3 sec. A number of the records are complicated also by longer period microseisms with periods of 1-2 sec and more. The appearance of the P wave on some of the seismograms ($\Delta > 2000$ km) is indistinct, being modulated by the short-period microseisms.

5. The Recorded Seismic Waves and Their Travel Times

In investigating the amplitudes and periods of seismic waves, in determining magnitudes, and in seeking solutions to a number of other problems discussed in the present paper, it is necessary to identify the waves.

Not all the seismograms show equally well-defined first arrivals. A sharp decrease in amplitude of the first arrival with increasing distance from the explosion site, against a background of noise, leads to the selection not of the first but of some subsequent arrival of the P wave. At nearby stations, where the magnification is extremely large, the leveling of the record strongly complicates interpretation of succeeding phases. At great distances, despite the higher sensitivity, it is sometimes difficult to distinguish a useful record, especially in the vicinity of the first arrivals, because of the background of microseisms.

The travel times for all arrivals that could be distinguished at the seismic stations in the special network, for each of the shots, are shown in Table 2. The directions of first arrivals, with rare exceptions, are in agreement with the data of American seismologists [32], as shown in Table 5 (third from last and last columns, p. 17).

Data on the travel times for the seismic waves from the Rainier shot are given in Table 3.

From the records at our disposal (chiefly from the special network of stations) of the American underground nuclear explosions—the Blanca, Logan, Tamalpais, and Rainier shots—we have constructed travel-time curves for the seismic waves traveling through the earth's crust, up to distances of 1100 km (Fig. 6).

These travel-time curves were constructed on the assumption that the waves distinguished are waves that were refracted at boundaries of the earth's crust consisting of granitic and basaltic layers. On this assumption straight lines were drawn through the experimental points on the graph. From the slope angle of these straight lines the apparent velocity of the corresponding wave was determined. And from the values of apparent velocity, in agreement with the previously adopted view of the geologic structure, the waves were interpreted in the following way.

At distances from 200 to 1000 km the points corresponding to first arrivals are closely approximated by straight lines, the slope angle of which gave a value of apparent velocity of 7.85 km/sec. This velocity is close to the velocity of the Pn wave, which is reflected at the Mohorovicic discontinuity. We shall speak of the characteristic peculiarities of the record of this wave in our discussion in the following section, in connection with correlating records by the shape of the curve.

After the relatively weak records of the Pn wave one may note prolonged and intense oscillations on the seismograms, belonging to a wave whose apparent velocity on the travel-time curves is 6.31 km/sec. This velocity corresponds to the longitudinal P* wave, refracted at the granite-basalt interface. This wave has been recorded among the first arrivals only at a distance of 100 km; at greater distances it is found among the succeeding later arrivals.

Transverse waves were distinguished only rarely. At but five stations could peaks be referred to the Sn wave, refracted at the Mohorovicic discontinuity. The apparent velocity of this wave is 4.55 km/sec. At the seismograph stations at distances 200-500 km from the source, the Sn wave gives a group of oscillations practically indistinguishable in amplitude and period from the neighboring segments of the record. After the Sn arrivals one may distinguish a wave traveling with an apparent velocity of 3.65 km/sec. This wave may have been refracted at the granite-basalt interface, the S* wave. According to data in [34], the transverse S* wave (or, more likely, the Lg wave) is found on the record up to distances of 2000 km from the source.

TABLE 3. Phases Distinguished on the Records of the Rainier Underground Nuclear Explosion

No.	Station	Δ, km	Arrival time h m s
1	Tinemaha	180.7	iP_n 17 00 29.7
2	Boulder City	182.2	iP_n 17 00 29.4
3	Hoover Dam	185.4	P_n 17 00 29.8
4	China Lake	197.5	P_n 17 00 31.2
			i 00 42.5
			iS_n 00 56.4
5	Eureka	256.7	iP_n 17 00 39.0
6	Woody	289.1	iP_n 17 00 43.2
7	Fresno	323.5	iP_n 17 00 48.1
8	Dalton	365.8	iP_n 17 00 53.1
			$iP*$ 01 01.2
			i 01 11.6
			$iS*$ 01 52.5
9	Riverside	370.8	iP_n 17 00 52.9
			$iP*$ 01 01.8
			i 01 05.5
10	King Ranch	379.8	P_n 17 00 55.1
11	Pasadena	382.2	iP_n 17 00 54.9
12	Reno	408.9	17 01 01.3
13	Palomar	430.4	P_n 17 01 01.0
14	Hamilton	482.8	P_n 17 01 07.1
15	Palo Alto	530.4	17 01 21.05
16	Berkeley	540.5	P_n 17 01 14.8
17	Salt Lake City	547.2	17 01 18.7
18	Mineral	582.6	17 01 46.9
19	Temporary station (from data in [4])	594.6	P_n 17 01 23.51
			$P*$ 01 40.52
20	Shasta	662.0	P_n 17 01 30.4
21	Tucson	736.7	eP_n 17 01 39.7
22	Boulder	1001.5	17 02 55.8
23	Laramie	1023.5	17 02 16.2
24	Butte	1025.4	17 02 00.4
25	Boseman	1034.7	17 02 50.5
26	Hungry Horse	1252.2	17 02 52.1
27	Rapid City	1337.2	17 02 51.0
28	Victoria	1388.4	17 03 13.1
29	Fayetteville	1968.0	P 17 04 11.5
30	College	3716.3	P 17 06 41.0

$*\varphi = 35°13'30.5"$ N; $\lambda = 110°02'03.2"$ W.

On Fig. 6 it may be seen that the first arrivals on the seismogram of the Arys' explosion (see § 10) fit fairly well on the travel-time curve for the Pn wave.

There are a number of papers by American authors [16, 23, 24] in which data are presented on the structure of the earth's crust and on the velocity of seismic waves in the region of Southern California and Nevada. Since some of the data by which, in the present paper, velocities were determined came from stations in precisely this region, it is of interest to compare the values of velocity obtained with the results of some other authors. In Table 4 we show some comparative data on the velocities; these attest to the correctness of interpretation of the waves distinguished on the seismograms, and they give one confidence in using the results obtained in further work.

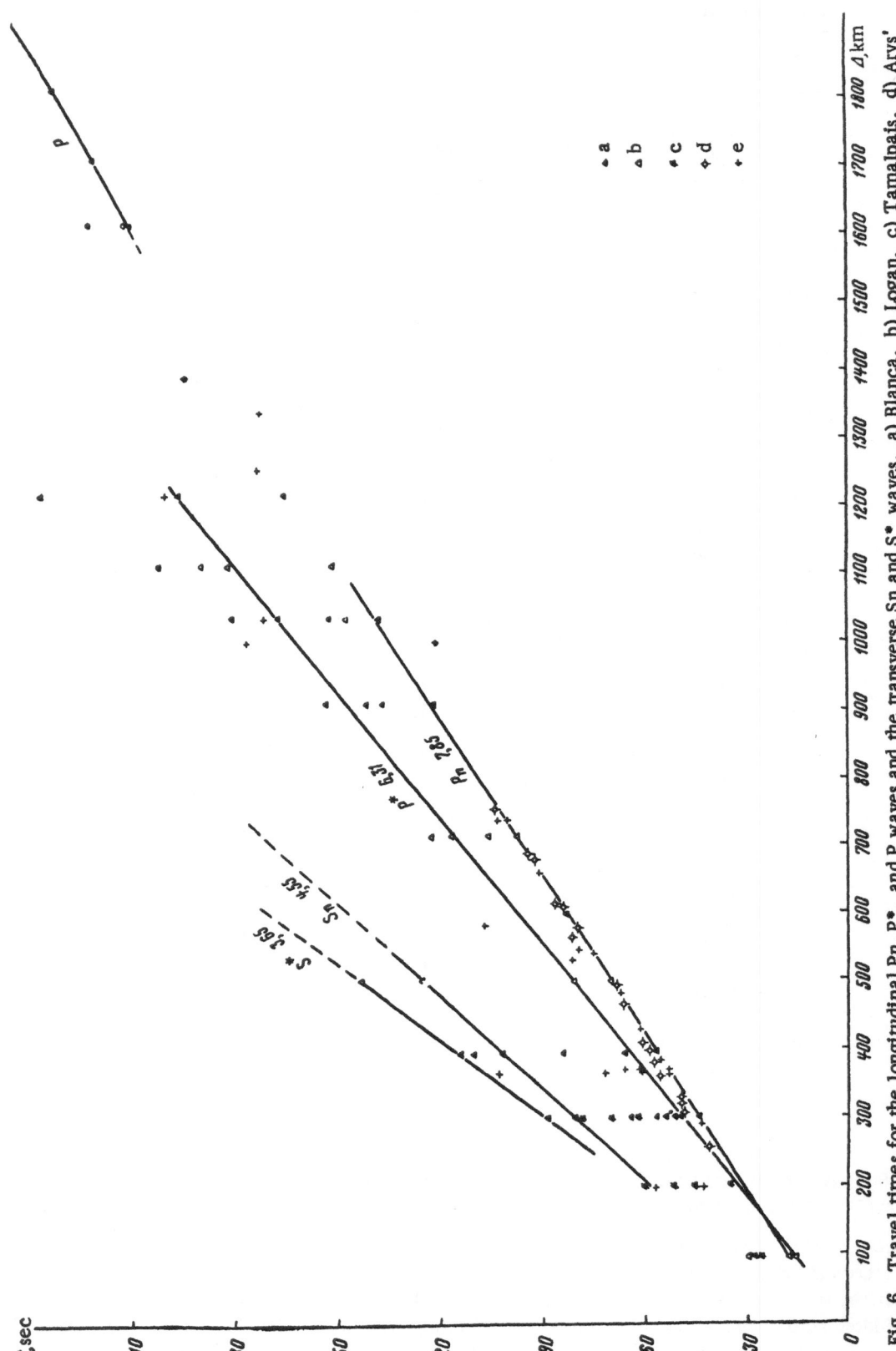

Fig. 6. Travel times for the longitudinal Pn, P*, and P waves and the transverse Sn and S* waves. a) Blanca, b) Logan, c) Tamalpais, d) Arys' explosion (conventional explosive), e) Rainier.

At distances of 1200-10,100 km from the source the first wave recorded is the P wave. The instants of arrival of this wave, taken directly from seismograms and published in seismological journals, are shown in Table 2. On the time-distance curve (Fig. 6) at distances greater than 1200 km we have plotted the arrival times of the P wave.

It should be noted that records of the P wave are generally absent on instruments with comparatively high magnification at the special stations located at distances approximately from 3000 to 4000 km from the source. The high

TABLE 4. Velocities of Seismic Waves

No.	Region	Reference	Velocity, km/sec				
			longitudinal wave			transverse wave	
			$\overline{P}$	$P*$	P_n	$S*$	S_n
1	Southern California	[23. 24]	6.0	6.5	8.1—8.2	3.75	4.6
2	Nevada-Southern California	[16]		6.1	8.0—8.2		
3	To the east of Nevada	Present paper		6.31	7.85	3.65	4.55

magnification of an instrument may be determined by the high level of the microseismic background. It is possible that the lack of record may be due simply to unfavorable locations or arrangements of the instruments.

On the records of the Benioff seismographs, the oscillations of the P wave at a distance of 2000 km from the source continued for more than two minutes (Fig. 5). On the records of the SVK-M seismograph, the oscillations of the P wave form, at most, several peaks, the maximum appearing at the second peak. At the Tiksi station the first phase recorded was rarefaction. At distances of 8,300 and 10,080 km the compressional phase (first peak) can scarcely be discerned (Figs. 7-11).

The KPK wave, which passes through the core, was recorded on SVK-M seismographs at the Soviet stations of Mirnyi and Banger Oasis (Antarctica), at distances greater than 16,000 km from the source.

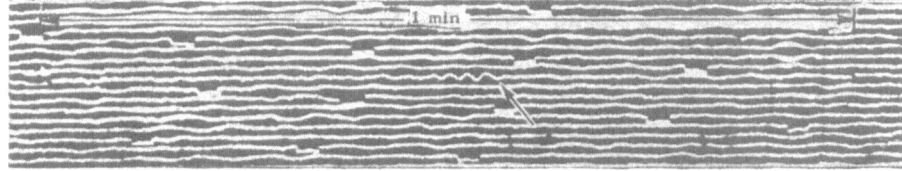

Fig. 7. Record of the Blanca shot at the Tiksi station. Δ = 6890 km, V = 24,000.

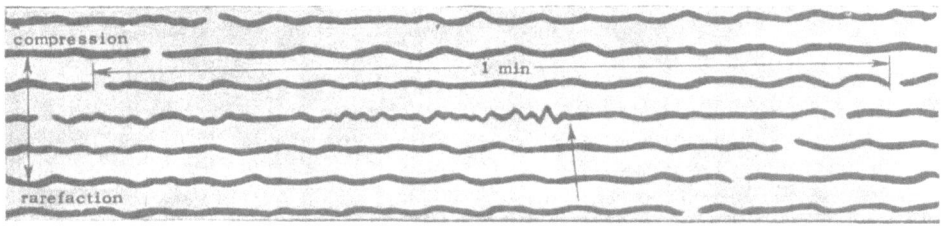

Fig. 8. Record of the Blanca shot. Δ = 8300 km, V = 23,000.

Thus, as seen from the Logan and Blanca underground nuclear explosions and also from a number of large TNT explosions, information on which is presented in § 10 of the present paper, it has been established experimentally that the Pn wave appears as the first arrival in continental regions at distances of 200 to 1000-1100 km from the source, whereas the P wave appears first at distances ranging from 1200 km to, approximately, 10,000 km (the Blanca shot).

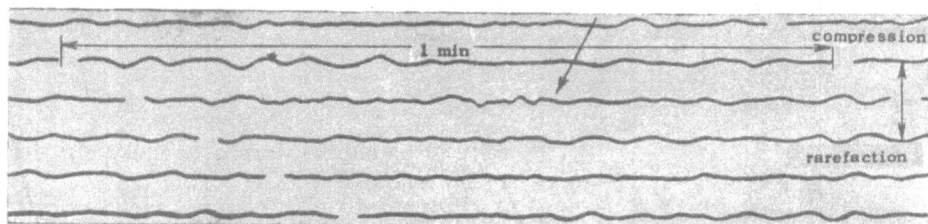

Fig. 9. Record of the Blanca shot at the Pruhonice station (Czechoslovakia). Δ = = 9180 km.

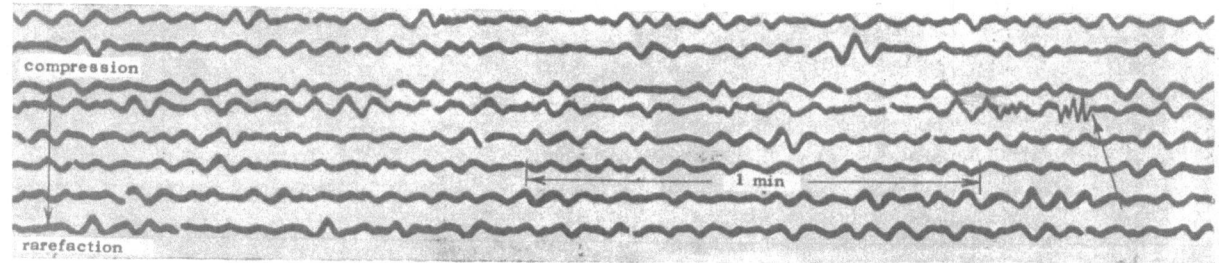

Fig. 10. Record of the Blanca shot. Δ = 10,080 km, V = 28,000.

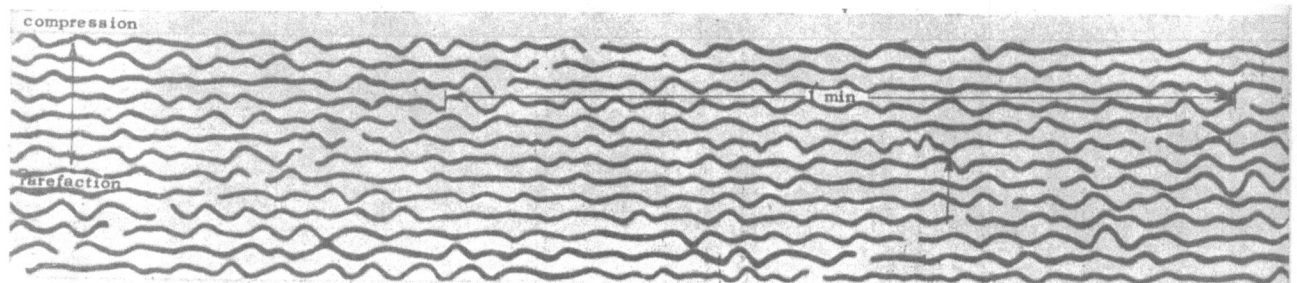

Fig. 11. Record of the Blanca shot at the Mirnyi station. Δ = 16,250 km, V = 12,000.

A transverse wave is observed in the succeeding arrivals at approximately the same distances as the Pn wave, but the records of this wave are indistinct. The Lg wave is found on records at distances up to 2000 km. On the seismograms at our disposal no transverse waves are recorded for the interval of 2000 to 10,080 km from the source; this may be due to the small amplitude of the waves, and possibly, also to less than optimum parameters of the recording apparatus.

6. Character of the First Arrivals on the Seismograms

It is known that the first arrival of a longitudinal wave from an explosion should be a compressional wave. Among the records of the underground nuclear explosions, at some of the stations, both in the special network and among the regularly operating stations, the first arrival proved to be a rarefactional wave. In Table 5 we have presented all the information from [32] relative to the direction of motion of the first arrivals; a compressional wave corresponds to a "+" sign, a rarefactional wave to a "−" sign, and a "?" sign signifies a lack of record or a doubtful record; no sign at all indicates that no information is available for the indicated station. In Table 5 we have assigned no number to the stations for which there is no information on the first arrivals or for which the coordinates are unknown. The distribution of signs at the stations of the special network is shown in Fig. 1 (the figures correspond to the sequential numbers in Table 5).

In order to discover the cause of the change in direction of the first motion, a correlation was made according to shape of the oscillation in the initial part of each record for different distances among the special-network stations.

TABLE 5. Direction of the First Motion of the Longitudinal Wave from the Blanca and Logan Underground Nuclear Explosions, Taken from [32]

No.	Name of station	Δ,km	Az°	i°	Blanca time of first arrival h m s	Blanca direction of motion	Logan time of first arrival h m s	Logan direction of motion
1	Beatty	59	175	67	15—00—10.4	+	06—00—10.5	+
	Camp Mercury	61				+		+
	Desert Rock	64			15—00—10.7	+	06—00— (07)	+
	Alamo-b	95			15—00—16.5	+	06—00—16.4	+
2	Alamo-a	96	95	67				+
	Tonopah	133			15—00—23.1	+		
	Lockes	140			15—00—23.6	+	06—00—23.7	+
3	Las Vegas	148	140	54	15—00—27	—	06—00—25.6	+
4	Tinemaha	181	270	54	15—00—30	+	06—00—30	+
5	Boulder City	181	5	54	15—00—29	+	06—00—30	+
6	Haiwee	195	230	54	15—00—32	+	06—00—32	+
7	China Lake	197	220	54	15—00—31	+	06—00—31	+
8	Hawthorne	203	300	54		+		+
9	Eureka	258	5	54	15—00—40	+	06—00—40	+
10	Isabella	267	230	54	15—00—40	+	06—00—40	+
11	Woody	289	235	54	15—00—44	+	06—00—44	+
12	Mt. Carmel	301	95	54		+	06—00—48	+
13	Fresno	323	260	54	15—00—48	+	06—00—48	+
14	Fort Tejon	353	225	54	15—00—52	—	06—00—52	+
15	Dalton	366	205	54	15—00—53	+	06—00—53	+
16	Riverside	371	200	54	15—00—54	+	06—00—53	+
17	King Ranch	380	240	54		+		+
18	Pasadena	382	210	54	15—00—55	+	06—00—55	+
19	Hayfield	389	175	54	15—00—55	+	06—00—56	+
20	Grand Canyon	395	115	54		+		
21	Reno	409	310	54	15—00—59	+	06—00—59	—
22	Palomar	430	190	54	15—01—01	+	06—01—01	+
23	Santa Barbara	439	230	54		+	06—01— (04)	?
24	Mt. Hamilton	483	275	54	15—01—08	+	06—01—08	+
25	Flagstaff	500	120	54		+		+
26	Barrett	503	185	54	15—01—10	+	06—01—10	?
27	Palo Alto	530	275	54	15—01—10	—		
	San Nicolas Island	531			15—01—13	+	06—01—13	—
28	Berkeley	541	280	54	15—01—16	+	06—01—16	+
29	Salt Lake City	548	40	54	15—01—19	+	06—01—19	?
30	San Francisco	557	280	54	15—01—22	?		—
31	Mineral	583	310	54	15—01—20	—	06—01—21	?
32	Holbrook	600	120	54		+		
33	Ukiah	649	290	54	15—01—47	—		
34	Shasta	662	305	54	15—01—30	+		
35	Gallup	715	115	54				—
36	Tucson	736	135	54	15—01—39	—	06—01—39	—
	Rifle	790			15—01—48	+	06—01—49	
	Albuquerque	809	115	54		?		

TABLE 5 (Cont'd)

No.	Name of station	Δ,km	Az°	i°	Blanca time of first arrival h m s	Blanca direction of motion	Logan time of first arrival h m s	Logan direction of motion
37	Boulder	1002	70	54	15—02—14	+	06—02—14	+
	Corvallis	1015				?		?
38	Laramie-a	1024	60	54	15—02—16	+	06—02—16	+
39	Butte	1027	15	54	15—02—20	+	06—02—20	?
40	Boseman	1036	25	54			06—02—22	+
41	Laramie-b	1036	60	54		+		?
	Tucumcari-a	1111						?
42	Tucumcari-b	1215	110	43		—		
43	Hungry Horse	1253	10	43	15—02—45	?	06—02—45	+
44	Canyon	1313	110	42				+
45	Rapid City	1338	50	42	15—02—55	—	06—02—53	—
46	Victoria	1390	335	42	15—03—13	+		
47	Tulia	1398	110	42		+		
48	Lawton	1610	95	41		+		+
49	Quanah	1707	100	41		..		
50	Tishomingo	1804	95	40				+
51	Dallas	1833	100	40		+		
52	Garland	1842	100	40		..		+
	Lawrence	1845			15—03—53	?		
53	Antlers	1902	95	39				+
54	Fayetteville	1968	85	38	15—04—09	+		+
55	Mena	2011	95	38		—		
56	Hot Springs	2111	95	38		+		?
	Little Rock	2162				?		
	Star Lake	2183					06—04—35	?
57	Salem	2209	80	37		—		
	East Merrick	2238						?
	Florrisant	2269			15—04—41	?		?
58	St. Louis	2282	80	35	15—04—41	?	06—04—42	+
59	Ste. Genevieve	2305	80	34				—
	Linko	2495					06—05—01	?
	Providence	2506						?
	New Albany	2665				?		
	Opasatika	2998			15—05—44	?		
	Cambridge	3017				?		?
	Chapel Hill	3309			15—06—09	?		
	Renaud	3309				?		
60	Ottawa	3474	65	26	15—06—23	+	06—06—24	+
	Forest City	3502						?
	Palisades	3650				?		?
61	Shawnigan Falls	3704	60	26	15—06—40	+		?
62	College	3717	335	26	15—06—42	+	06—06—42	+
63	Seven Falls	3852	60	25		—	06—06—56	—
64	Bangor	4021	65	25		—		?

Correlations for the first peaks from the Blanca shot are shown in Fig. 12a, from the Logan shot in Fig. 13a; correlations for the third peaks, from the same shots, are shown in Figs. 12b and 13b respectively. If the rarefactional waves were due to the nature of the source, then there should be some correlation in the records in comparing the shapes of the first peaks; but there is no such correlation in Figs. 12a and 13a. Figures 12b and 13b clearly show that a good correspondence in shape of the record may be obtained only when the third peaks are compared. Consequently, the change in sign is due to the fact that the first arrival and, in many records, even the second arrival are not

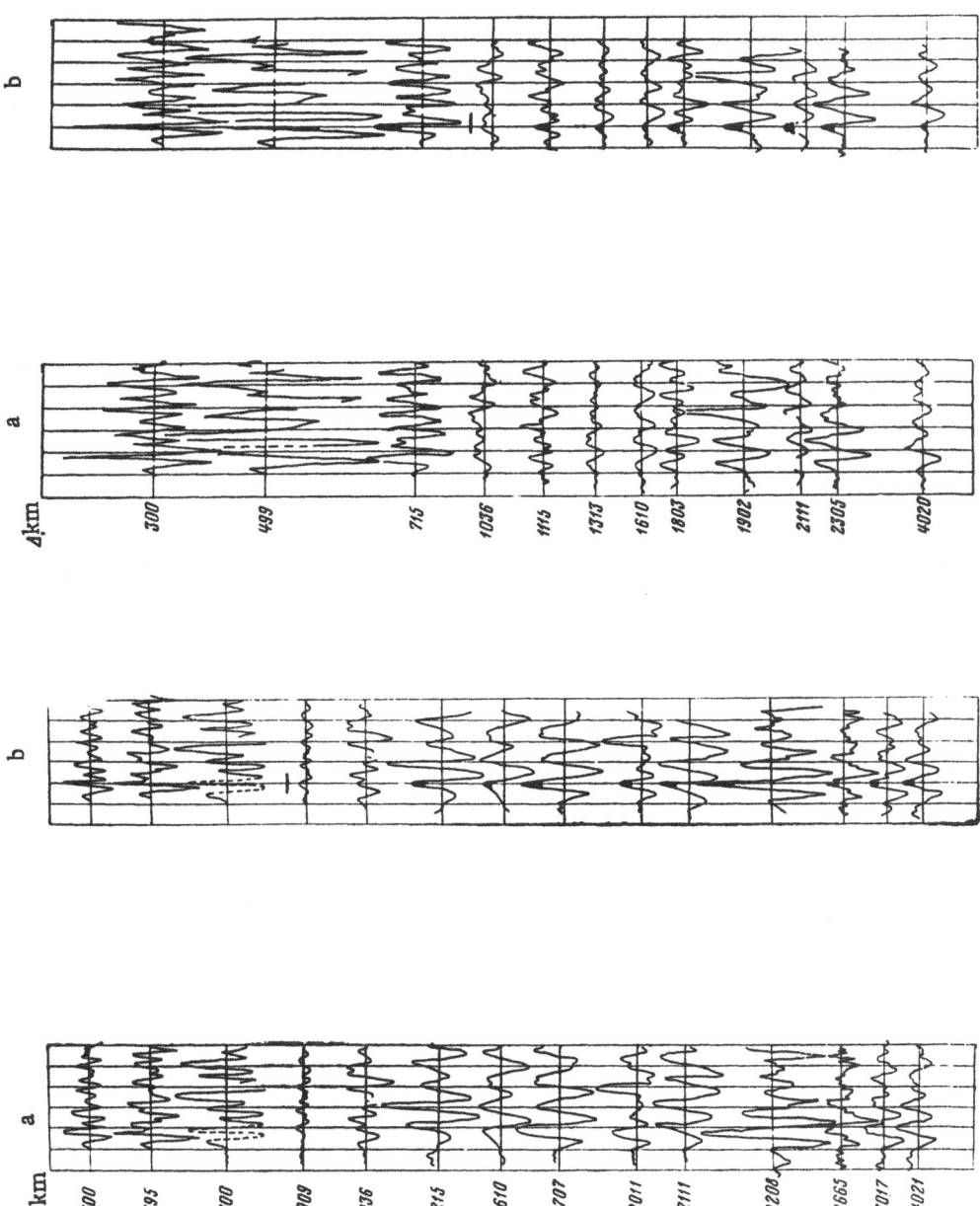

Fig. 12. Correlation of initial parts of record for the Blanca shot. a) For the first arrival; b) for the third peak.

Fig. 13. Correlation of the initial parts of record of the Logan shot. a) For the first arrival; b) for the third peak.

distinguished because of the background of noise. The inability to make correlations for the distance of about 1000 km from the source points to a change of waves here: the Pn wave is replaced by the P wave.

Thus, we may note that one peculiarity of the seismic records of underground explosions obtained on a Benioff seismograph is the comparatively low intensity of the compressional phase. This peculiarity may be associated with damping of high-frequency oscillations and with the distorting effect of the recording apparatus.

In the Benioff instruments the ratio of amplitudes of the compressional and rarefaction phases for the Blanca shot amounts to 0.3 on the average at a distance from the source of about 700 km. The ratio between amplitude of the first motion and the maximum amplitude is approximately the same. At greater distances from the source this ratio decreases somewhat.

It may be expected that when apparatus with optimum parameters is used, leading to less distortion of the first arrival, and when a large number of seismographs are used in a group, thus lowering the noise level, the first arrivals on the records will be clear. At two sites about 1600 km from the source [34], groups of two and of four seismographs were used, coupled in each arrangement with a single galvanometer. With these arrangements the signal-to-noise ratio increased approximately $\sqrt{n}$ times, where $\underline{n}$ represents the number of seismographs in a group.

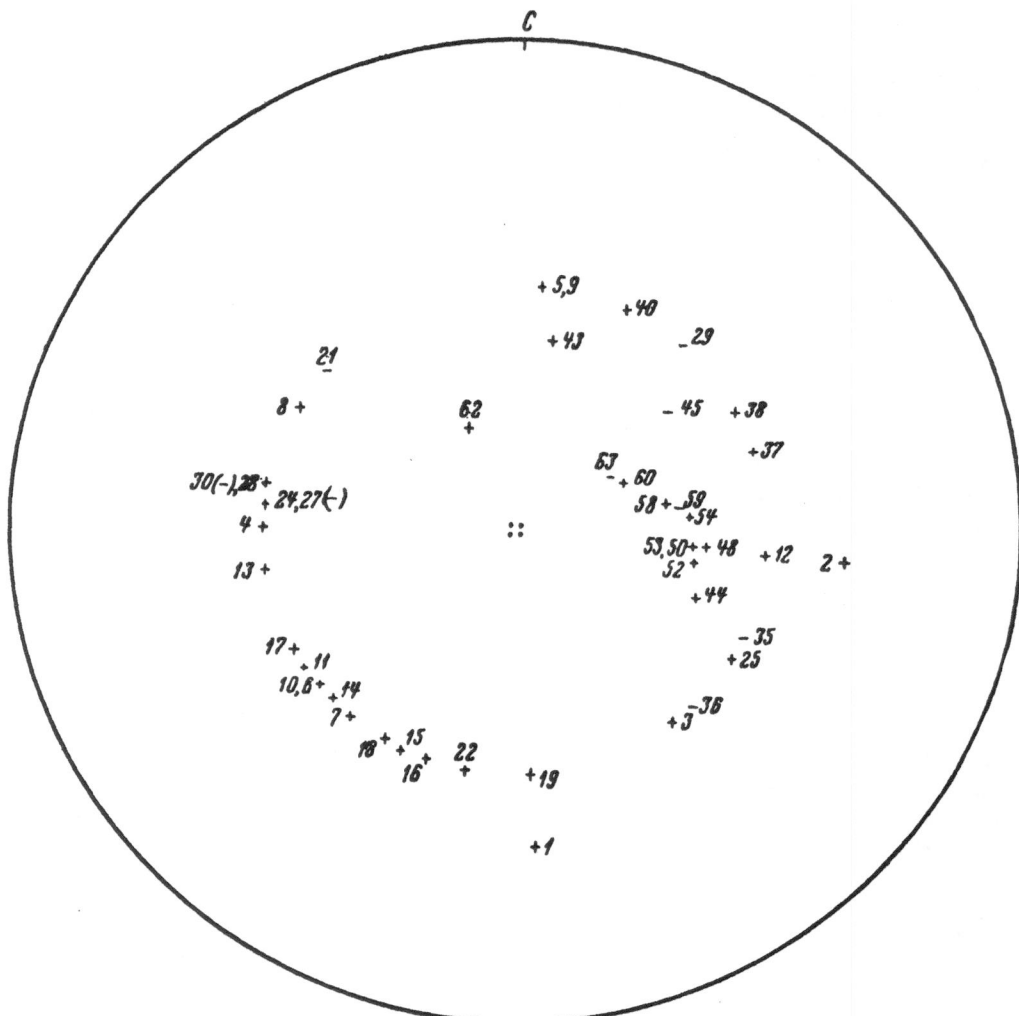

Fig. 14. Distribution of positive and negative directions of first arrivals on a Wulff net, for the Logan shot.

In analyzing the distribution of signs (directions) of first motions on the records, a method of treating the observations of first arrivals was used such as is used in studying the mechanism of earthquakes [3]. All existing data from permanent and temporary stations in the USA and in Canada concerning the direction of first motion of the Pn and P waves was plotted on a Wulff net (Logan shot, Fig. 14; Blanca shot, Fig. 15). In Table 5 the azimuths from source to station are shown with a precision of 5°; the incident angle $\underline{i}$ of the P wave is given by the travel-time curve of Jeffreys-Bullen [3].

At distances approximately up to 100 km the first arrival was assumed to be the P* wave, at distances approximately up to 1000 km, the Pn wave; the angles of total internal reflection were determined on the basis of the velocities v_P = 6.0 km/sec, v_{P*} = 6.5 km/sec, and v_{Pn} = 8.1 km/sec (Table 4). On Figs 14 and 15 the "+" signs represents a compressional wave, the "−" sign a rarefactional wave, and the figures indicate stations corresponding to the sequential numbers in Table 5. If the coordinates of two stations (Az, i) coincide on the Wulff net, then two figures are placed side by side; if the movements at the stations are different, then the sign of the movement at the more distant station is placed in parenthesis beside the figure. Different data have been cited, apparently, for some

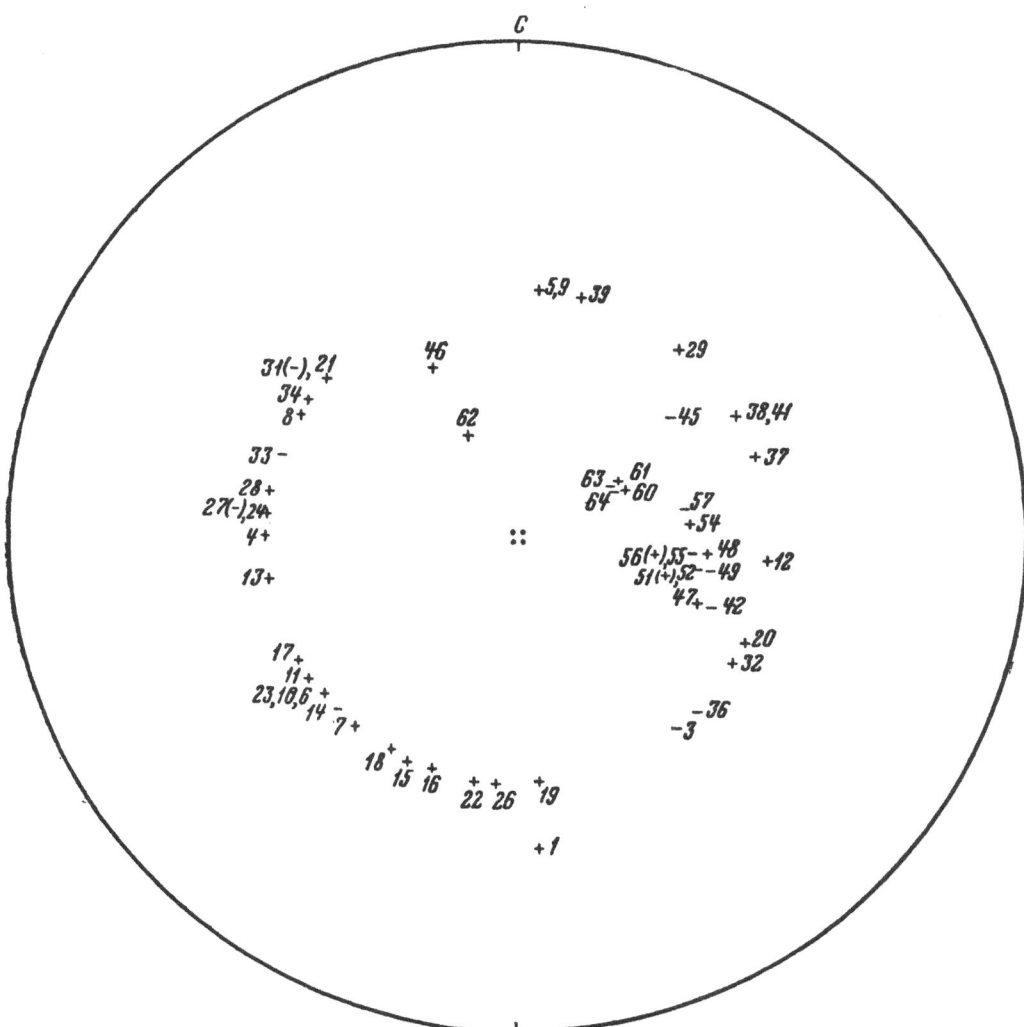

Fig. 15. Distribution of positive and negative directions of first arrivals on a Wulff net, for the Blanca shot.

of the indistinct arrivals as shown in various tables sent us by American seismologists. For example, one table gives a "−" sign at the Tucson station, another gives "?" (indicating an indistinct arrival); we have adopted the "−" sign in such cases.

The distribution of positive and negative directions of movement, as shown on Figs. 14 and 15, does not permit us to draw a nodal line separating the points of positive movement from points of negative movement. A study of the mechanism of actual earthquakes indicates that such a distribution is impossible for the great majority of quakes.

Consequently, the presence of rarefactional waves at individual stations is due not to the source but rather to the loss of the first arrival, a fact that may be detected, as has been shown, by means of phase correlation; the appearance of the rarefactional waves may also be due to an error in coupling the instrument. Thus, a composite study of the directions of motion of the first arrivals at all the stations and correlation of phases of these arrivals substantially increases the reliability of evaluating the first arrivals.

7. Periods of the Seismic Waves Recorded from Underground Explosions and Weak Earthquakes

A study of the periods of the seismic waves arising during explosions is necessary, both in order to select recording apparatus with the optimum frequency characteristics and in order to expose the features that may permit one to distinguish the record of an underground explosion from the record of an earthquake.

From the material on underground nuclear explosions we now have, we have been fully successful in determine the periods of body waves only: the longitudinal Pn, P*, and P waves and the transverse Sn and S* waves.

All the values of periods that are pointed out in the remainder of this paper (Tables 6 and 7) are observable on the record by the prevailing periods with maximum amplitudes for a given wave.

A comparison of periods recorded at different distances is necessary when records are obtained by a single type of instrument or by instruments with similar parameters.

The relationship between periods of different waves to distance from source is shown in Figs. 16-20.

At distances ranging from 200 to 1000 km, the periods of the Pn and P* waves recorded on Benioff seismographs range from 0.5 to 0.8 sec; individual values that dropped sharply were also observed.

The periods of the Pn and P* waves recorded at one station and for a single shot were the same, within the limits of measurement, although one may note a small increase (0.05-0.1 sec) in the period of the P* wave over the period of the Pn wave.

Extensive variation in the data for the Rainier shot (see the plots for numbers 5 and 7 on Fig. 16) is probably associated with errors in measurement because of a slow recording rate and because of indistinct records. Furthermore, most of the records of this shot were not made on Benioff vertical seismographs.

The periods of the Sn and S* waves at distances from 200 to 500 km change from 0.6 to 1.2 sec (Fig. 17).

The period of the Sn wave is a little larger than the period of the Pn wave, the difference increasing as the distance becomes greater.

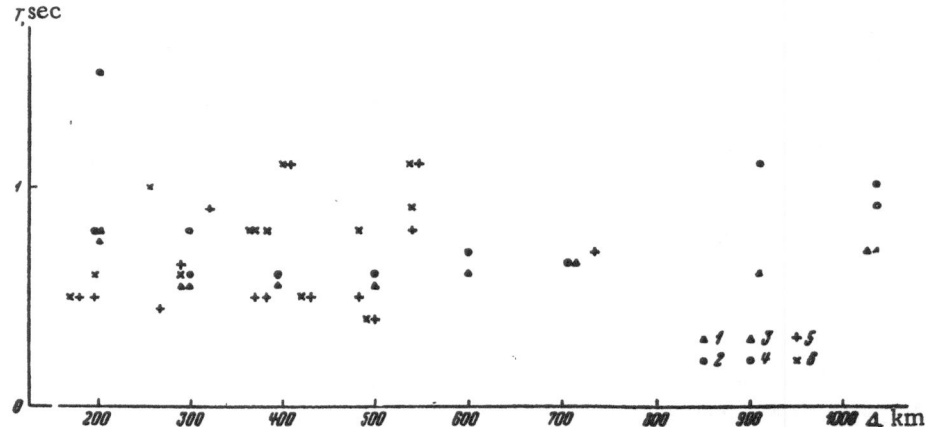

Fig. 16. Relationship between periods of Pn and P* waves and distance from site of underground nuclear explosions. 1) Pn, Blanca; 2) P*, Blanca; 3) Pn, Logan; 4) P*, Logan; 5) Pn, Rainier; 6) P*, Rainier.

At a distance of about 1000 km, where the Pn waves gives way to the P wave, the period of the longitudinal wave increases. For a period of Tp considerable variation is noted (from 0.7 to 1.2 sec) at distances between 1000 and 3000 km. Observations at great distances indicate values at the upper limit of the indicated variation (1.2- 1.4 sec); moreover, these data were obtained on a different instrument (SVK-M seismograph).

Within the limits of precision attained (0.1-0.4 sec), the periods of body waves seen on the records do not depend on the yield of the blast at distances as little as 200 km from the site. It might be thought that such a relationship exists for the high-frequency part of the spectrum on records taken near the source.

The few data concerning periods of surface waves from nuclear explosions indicate that the narrow-band apparatus records these waves with periods only slightly greater than the periods of the transverse waves. At distances of 100-500 km the period T changes from 0.8 to 1.2 sec. At greater distances the period increases to 1.5-2 sec. With instruments having a wider band of reception, surface waves with periods of 2-3 sec have been noted. Surface Rayleigh and Love waves, with periods of 7-10 sec, have been observed on records of long-period seismographs at Pasadena (380 km), Berkeley (540 km), and Palisades (3650 km).

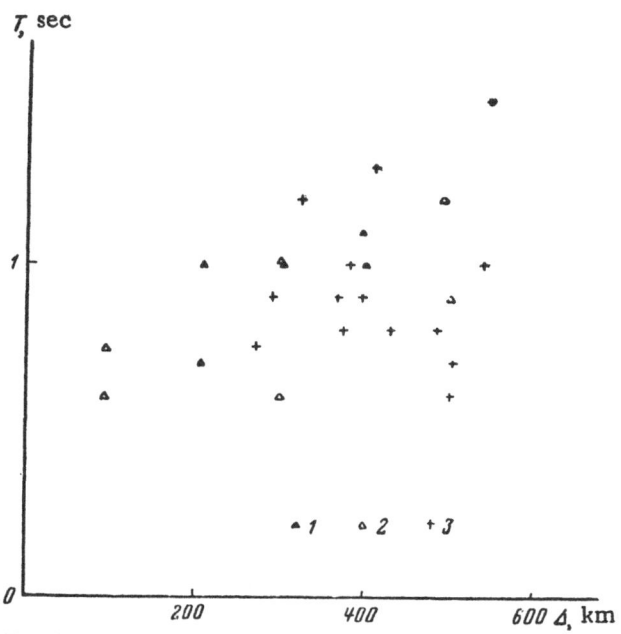

Fig. 17. Relationship between periods of Sn and S* waves and distance from site of underground nuclear explosions. 1) Blanca, 2) Logan, 3) Rainier.

An examination of the differences in frequency spectra of seismic waves from underground explosions and from earthquakes is important in seeking a way to recognize the record of an explosion from earthquake records. We have records from SVK, SGK, SVKM, and SGKM instruments of powerful blasts of chemical explosives (Fig. 21, 22), which may be compared with the records of earthquakes. Distance-period relationships have been obtained only for surface waves for earthquakes recorded by SVK and SGK seismographs [12]. In order to obtain data on the periods of body waves, records of earthquakes from the Frunze station for the period from January to June 1956 have been studied. The periods of longitudinal, transverse, and surface waves have been plotted on, the graphs on Figs. 18-20, against distance for nearby earthquakes and for the Arys' explosion of ordinary explosives, set off on December 19, 1957 (see § 10).

The periods of the longitudinal wave (Fig. 18) from earthquakes with a magnitude $M_s \approx$ 3-5 changes from 0.6 to 2.5 sec in the distance interval from 100 to 1000 km. The longitudinal wave from the explosion changed in period from 0.2 to 0.8 sec in the same distance interval. Consequently, although there is a difference in periods of longitudinal waves from earthquakes and from explosions, this criterion cannot always be used to distinguish explosions, since the values of the periods overlap.

The periods of the transverse wave (Fig. 19) change approximately from 1 to 4 sec at distances ranging from 100 to 1000 km. It was not possible to compare these values with data from explosions for any extensive interval of distance because of the difficulty of distinguishing transverse waves on most of the records of explosions obtained on SVK and SGK instruments. The periods of direct transverse S waves on the records of SGK-M seismographs were near 1.5 sec.

The difference in periods for surface waves from earthquakes and from explosions are more noticeable on the records of wide-range instruments [6]. In Fig. 20 the curve showing the period-distance relationship for surface waves from earthquakes is expressed in agreement with the data of Solov'ev and Shebalin [12] by the formula $T \approx 0.85\sqrt[3]{\Delta}$. Observations at the Frunze station are in accord with this relationship.

From the cited data on periods of earthquake waves it is impossible to establish any relationship between periods of longitudinal, transverse, and surface waves, on the one hand, and values of M_s and depth of focus, on the other, since the number of data treated for this purpose is too small.

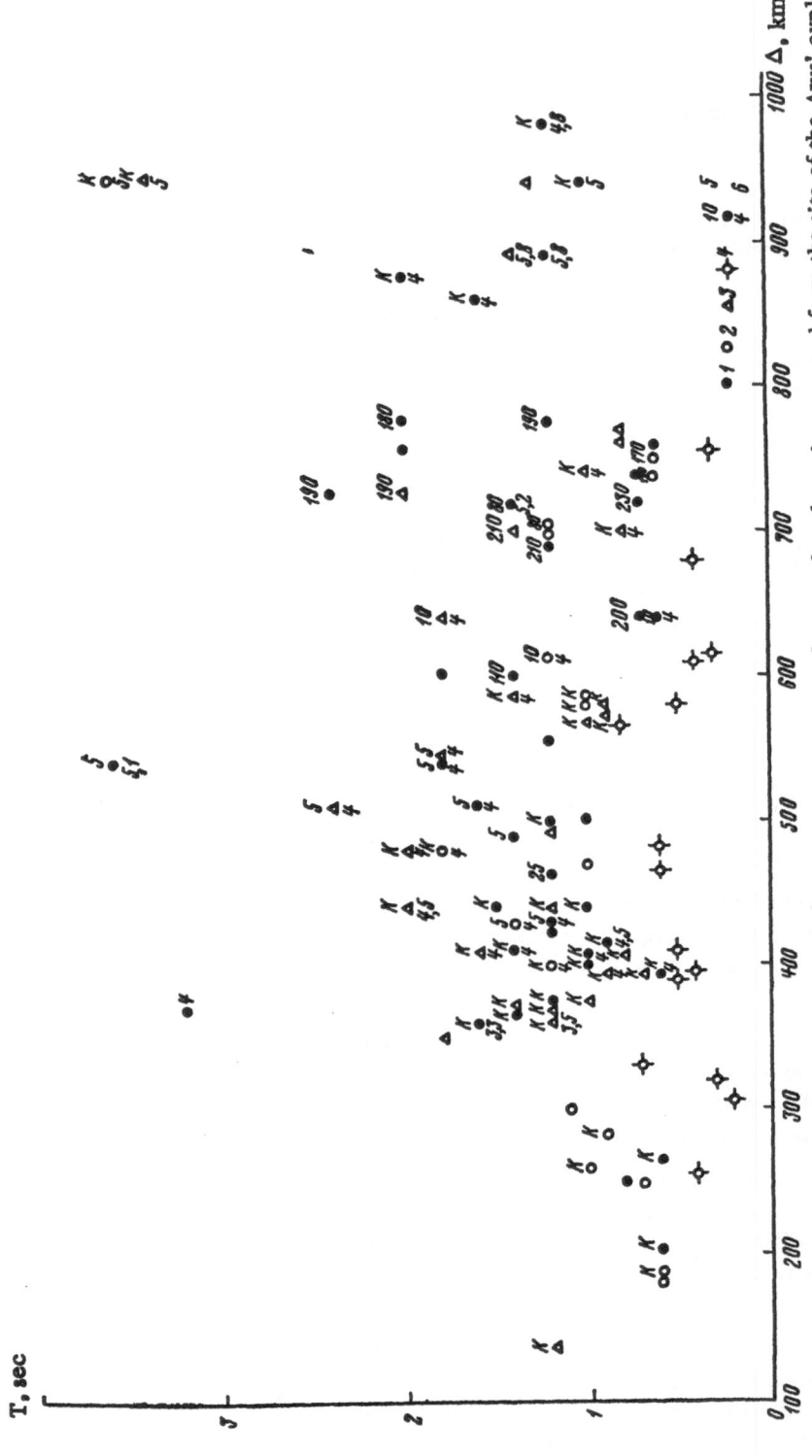

Fig. 18. Relationship between periods of longitudinal waves and distances from earthquake epicenters and from the site of the Arys' explosion. 1) Pn, 2) $\overline{P}$, 3) P*, 4) Pn for the Arys' blast, 5) depth to focus (K represents focus in the earth's crust), 6) value of M_S.

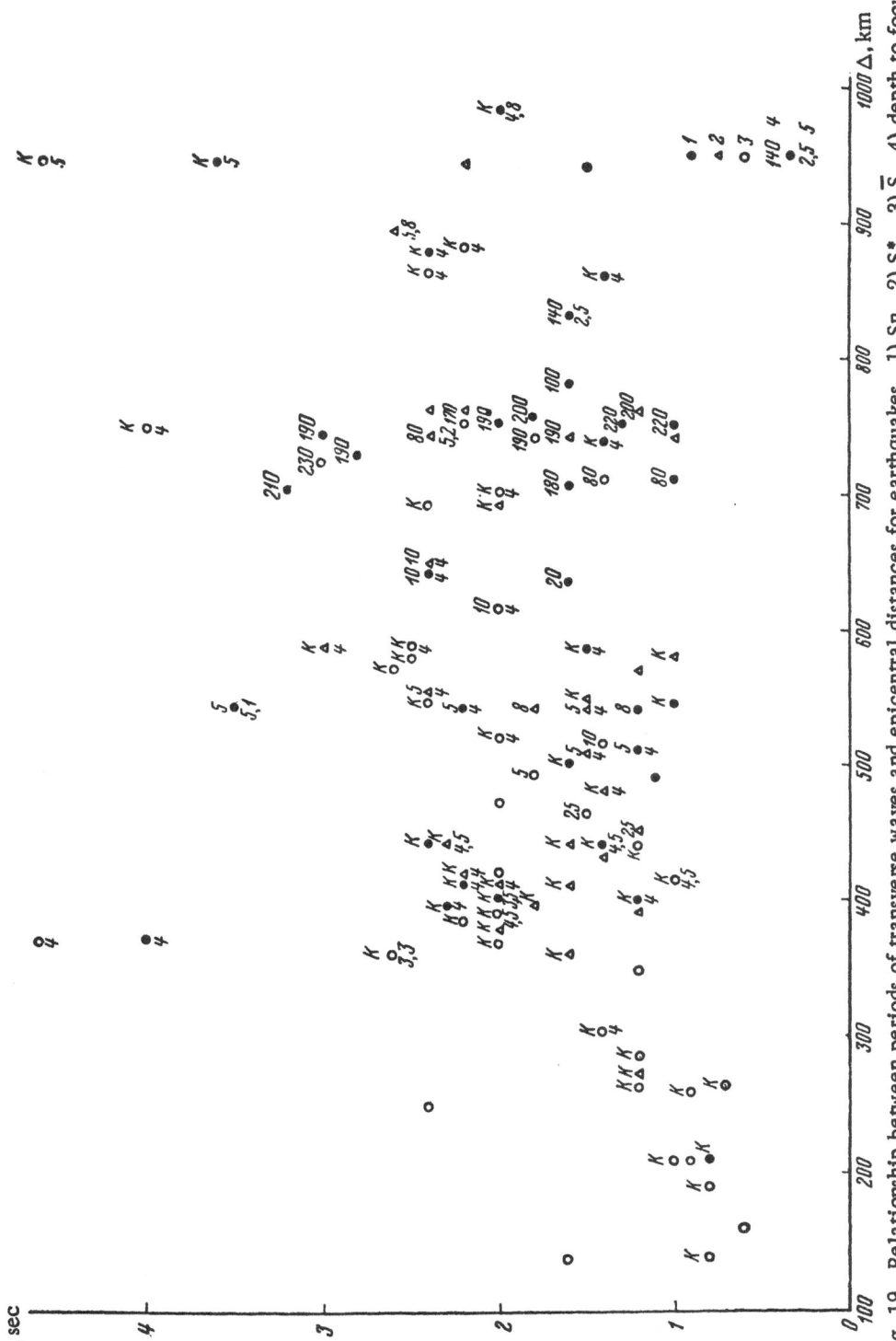

Fig. 19. Relationship between periods of transverse waves and epicentral distances for earthquakes. 1) Sn, 2) S*, 3) S̄, 4) depth to focus (K represents focus in the earth's crust), 5) value of M_S.

TABLE 6. Periods of Body and Surface Waves Recorded from Underground Nuclear

No.	Station	Δ, km	T_{P_n}, sec				T_P^*, sec			
			B	L	T	R	B	L	T	R
1		96.2			0.7				0.75	
2	Tinemaha	180.7				0.5				0.5
3	China Lake	197.5				0.5				0.6
4		203.4	0.8				0.8			
5		203.5		0.75	0.75				0.65	
6	Eureka	256.7								1.0
7	Isabella	267.0				0.45				
8	Woody	289.1				0.65				0.6
9		300.6	0.55	0.55	0.6		0.6	0.8	0.55	
10	Fresno	323.5				0.9				
11	Dalton	365.8								0.8
12	Riverside	370.8				0.5				0.8
13	King Ranch	379.8				0.5				0.5
14	Pasadena	382.2				0.5				0.8
15		395.1	0.55				0.6			
16	Reno	408.9				1.1				1.1
17	Palomar	430.4				0.5				0.5
18	Mt. Hamilton	482.8				0.5				0.8
19		498.9		0.55				0.6		
20	Hayfield	500.0				0.4				0.4
21	Barrett	502.8								
22	Berkeley	540.5				0.8				0.9
23	Salt Lake City	547.2				1.1				1.1
24		600	0.6				0.7			
25		714.5		0.65				0.65		
26	Tucson	736.0				0.7				
27		909	0.6				1.1			
28	Boseman	1036								
29		1036	0.7	0.7			1.0	0.9		
30		1111.5		0.7				1.2		
31		1215	1.0				1.0			
32	Hungry Horse	1253.5								
33	Fayetteville	1968.2								

*Initials are used to designate the explosions: B for Blanca, L for Logan, T for Tamalpais,

The period of the surface wave on records of the Arys' explosion (see Fig. 20 and Table 11) at distances of 300 to 1100 km is approximately 2 ± 0.5 sec; i.e., it is practically invariant with increase in distance. The same periods for surface waves were also measured for the explosion at Pokrovsk-Ural'skii (see Table 12 and § 10). Thus, at a distance of about 1000 km the period of the surface wave from an explosion will be approximately one-fourth the period of a surface wave from an earthquake. Such a difference in values may be used as a reliable criterion for distinguishing the record of an explosion from an earthquake record.

On the basis of the observed data, the concept has developed that the high-frequency spectrum of waves from explosions is due to the small dimensions of the source and to the short duration of its action. This view has been confirmed by the theoretical computations of V. I. Keilis-Borok [5], according to which, because of differences in

Explosions at Distances up to 2000 km *

T_{S_n}, sec			T_S, sec		T_L, sec			Type of seismograph
B	L	R	B	L	B	L	R	
	0.6			0.75	0.7—0.8			Benioff
		0.8					0.9	Sprengnether
							1.15	"
0.7					0.7—0.9			Benioff
			1.0					"
								Unknown
		0.75					1.0	Sprengnether
		0.9					0.95	"
1.0	0.6		1.0	1.0	1.0			Benioff
		1.2					1.4	Sprengnether
		0.9						Unknown
		0.8					1.0	Sprengnether
		1.0					1.0	"
1.1		0.9	1.0		1.0			Benioff
		1.3					1.6	Sprengnether
		0.8					1.0	"
		0.8					1.2	Benioff
	0.9			1.2	0.8—1.2			"
		0.6						Sprengnether
		0.7						"
		1.0						Benioff
						3		Galitzin
						7		"
		1.5				2	1.6	Unknown
								Benioff
								"
						1.5		Unknown
					2			Long-period Benioff
						2		Unknown
								Benioff
								"
								"
					3			Long-period
						3		Unknown

and R for Rainier.

size of focus, surface waves from explosions should have a smaller period than surface waves from earthquakes, if the explosions and the earthquakes are comparable in the amount of energy given off as elastic seismic waves.

The longitudinal waves from underground nuclear explosions were recorded on SVK-M seismographs at distances ranging from 7000 to 10,000 km with periods ranging from 1.2 to 1.4 sec; the longitudinal wave from the Arys' explosion was recorded at a distance of 2600 km with a period of 0.7 sec. Similar periods for longitudinal waves were obtained on a number of records of earthquakes by these same instruments.

Note. The fact that the periods of surface waves from underground explosions change little with distance may be due to the very narrow spectrum of these waves. The views here proposed are supported by the presence of a long train in the record of the surface wave.

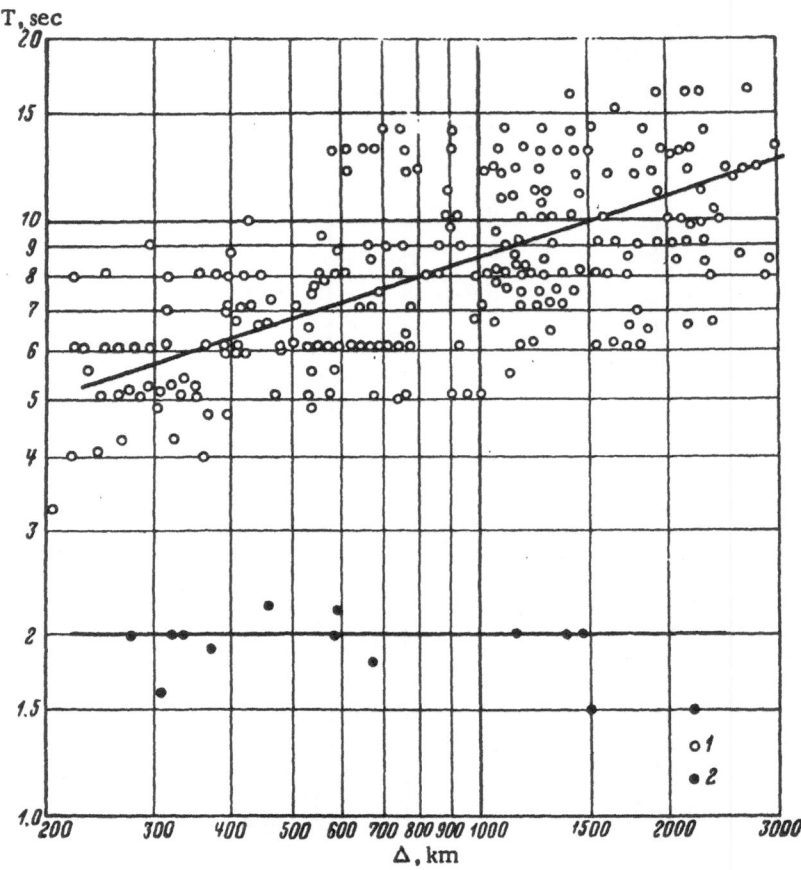

Fig. 20. Relationship between periods of surface waves and distances from earthquake epicenters and from the site of the Arys' explosion. 1) Period for earthquake, 2) period for Arys' explosion.

TABLE 7. Periods of P Waves Recorded from Underground Nuclear Explosions

No.	Station	Δ,km	T_p, sec		Type of seismograph
			Blanca	Logan	
1		1313.1		0.9	Benioff
2		1610	1.1	1.1	"
3		1707	1.1		"
4		1803.7		0.7	"
5		1902.1		1.0	"
6		2011	0.9		"
7		2111		1.0	"
8		2111.3	1.2		"
9		2208	0.8		"
10		2305		1.1	"
11		2506		0.7	"
12		2665	1.2		"
13		3017	1.2		"
14		4021	1.1		"
15	Tiksi	6890	1.2	1.2	SVK-M
16	Temporary station	8330	1.4		"
17	Temporary station	10080	1.2		"
18	Mirnyi	16250	RKR 1.7		"

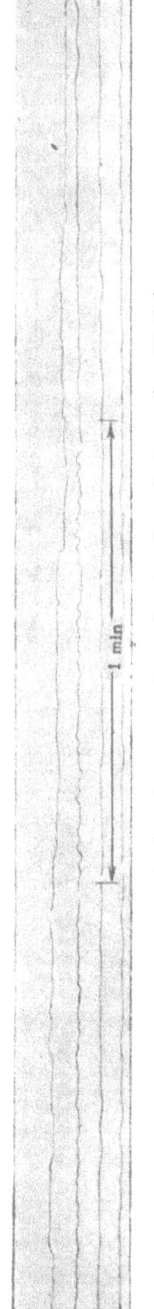

Fig. 21. Record of surface wave from the Arys' explosion. Δ = 1140 km.

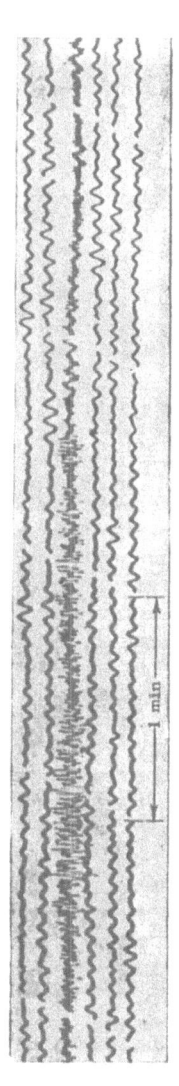

Fig. 22. Record of surface wave from explosion at Pokrovsk–Ural'skii. Δ = 1390 km.

From prints of records of the Logan and Blanca underground nuclear explosions and from seismograms made at Soviet seismic stations recording these same explosions, the amplitudes of the longitudinal Pn and P waves have been measured, and the ground displacements of these waves have been plotted as a function of distance.

The amplitude of the Pn wave was measured on seismograms of the Logan and Blanca shots for distances from 200 to 1100 km, and for the P wave, for distances from 1200 to 10,080 km. The amplitudes of the first, second, and third peaks were measured for both Pn and P waves where this was possible; the third peak generally represented the maximum.

In computing amplitudes and periods of ground displacement, no corrections were made for the pulsating (unsteady) nature of the oscillations in the first and succeeding kicks on the record. This correction, as is well known, has considerable value for the first peak. On the records of Benioff seismographs this leads to a considerable decrease in amplitude of the first kick. For succeeding peaks, especially for the maximum, this correction is generally insignificant, and the measured values are near the actual values.

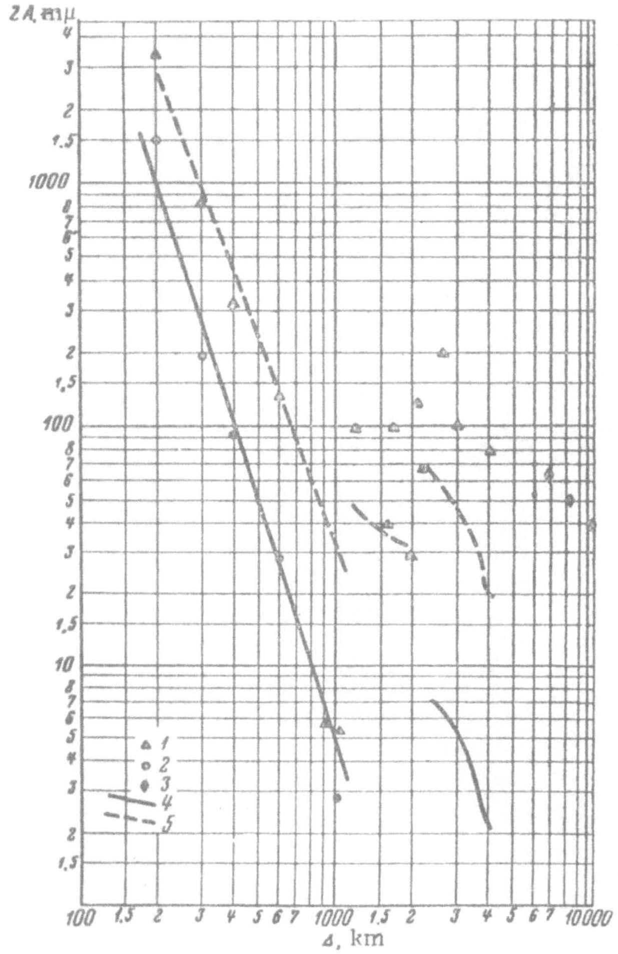

It should be noted that, apparently, the relative distortion of amplitudes and periods is approximately uniform for all distances because of the use of identical instruments along the profile. Therefore, in determining the character of the changes in amplitude with distance, distortions introduced in the record by the pulsating nature of the oscillation cannot be considered.

The amplitudes of the Pn and P waves, with designations of which peak was measured, are shown in Table 8. The nature of the decrease in amplitude of the Pn and P waves for the Tamalpais, Logan, and Blanca shots is illustrated in Figs. 23-25. The curves in these figures are plotted with coordinates of log A and log Δ, where A is amplitude in millimicrons and Δ is the epicentral distance in kilometers. Experimental values of doubled amplitude of the first and maximum peaks are plotted on the figures, for the Blanca shot (Fig. 23) and for the Logan and Tamalpais shots (Fig. 24).

It should be noted that the amplitude of the Pn wave (3μ) for the Rainier shot at a distance of 594.6 km, as indicated in [20], was incorrectly determined: the reported value is approximately one hundred times the actual value for the more powerful Blanca shot.

The lines on the graph averaging the experimental Pn amplitudes for corresponding peaks from the Logan and Blanca nuclear shots are similar; they are straight and have a common slope.

From an examination of Figs. 23 and 24 it may be seen that the slopes of the straight lines averaging the values of the first and maximum peaks of the Pn wave from the Logan and Blanca shots diverge slightly. The lines averaging the values of the first peaks dip more steeply with increasing epicentral distance than do the

Fig. 23. Relationship between amplitudes of Pn and P waves and the epicentral distance for the Blanca shot. 1) For maximum peak; 2) for first arrival; 3) for maximum peak on records of the SVK-M seismograph; 4 and 5) lines transferred from the graph of Appendix V in paper [14].

TABLE 8. Relationship between Displacements of the Pn and P Waves (in mμ) and the Epicentral Distance for Explosions of the Hardtack II Series Designations: A_1 is amplitude of first motion; $\frac{1}{2} A_{23}$ is the average amplitude between the second and third peaks; A_{max} is the amplitude of the maximum peak; T_{max} is the period of the oscillation of the maximum peak; $V_{T_{max}}$ is the magnification corresponding to T_{max}

No.	Δ km	Blanca A_1	Blanca $\frac{1}{2}A_{23}$	Blanca A_{max}	Blanca T_{max}	Blanca $V_{T_{max}}$	Logan A_1	Logan $\frac{1}{2}A_{23}$	Logan A_{max}	Logan T_{max}	Logan $V_{T_{max}}$	Tamalpais A_1	Tamalpais $\frac{1}{2}A_{23}$	Tamalpais A_{max}	Tamalpais T_{max}	Tamalpais $V_{T_{max}}$
1	203.4	750	1686	1686	0.8	1600	250	835	835	0.75	14 000	3.7	19.2	19.2	0.75	110 000
2	203.5	98	410	410	0.55	6100	40.8	221	221	0.55	29 400	1.3	4.8	4.8	0.6	232 000
3	300.6	46	161	161	0.55	26 000	7.1	41.6	41.6	0.55	282 000					
4	395.1															
5	498.9	14	66.5	66.5	0.6	58 600	1.9	13.9	13.9	0.65	208 000					
6	600															
7	714.5	—	2.9	2.9	0.6	173 000	—	1.6	1.6	0.7	517 000					
8	909	1.4	2.7	2.7	0.7	365 000	—	5.2	5.2	0.7	217 000					
9	1036															
10	1111.5															
11	1215	—	22.3	48.7	1.0	98 500	—	9.5	13.5	0.9	52 000					
12	1313															
13	1610	—	10.9	19.7	1.1	124 000	—	3.4	12.3	1.1	89 500					
14	1707	—	39.5	48.9	1.1	87 500	—	2.6	4.2	0.7	334 000					
15	1803															
16	1902															
17	2011	—	10.4	14.6	0.9	192 000	—	8.7	24.4	1.0	172 000					
18	2111	—	43.5	62.3	1.2	57 600	—	7.5	14.8	1.0	88 000					
19	2208	—	32.5	34.0	0.8	197 000	—	21.0	58.0	1.1	52 000					
20	2305															
21	2506	—	64.0	98.8	1.2	27 300	—	14.0	—	0.7	15 700					
22	2665	—	51.5	51.8	1.2	38 600										
23	3017	—	40.0	40.7	1.1	51 500										
24	4021															
25	6890	Tiksi		32.0	1.2	25 000			16.0	1.2	25 000					
26	8300	Temporary station		26.0	1.4	19 000										
27	10080	Temporary station		20.0	1.2	30 000										

lines averaging the maximum peaks. This phenomenon may be due to high absorption of the short-period components abundant in the first arrival.

From [9] it is known that the dimunution of the head waves occurs according to the law

$$A_i = A_0 \left(\frac{\Delta_i}{\Delta_0} \right)^{-n} e^{-\alpha(\Delta_i - \Delta_0)}. \tag{1}$$

Here A_0 and Δ_0 are the amplitude and epicentral distance respectively at the point of observation, situated at the initial point of the head wave; A_i and Δ_i are the amplitude and epicentral distance at a point of observation beyond Δ_0; and $\underline{n}$ and α are, respectively, the exponent of the scattering function and the amplitude coefficient of absorption of the head waves. The average value of $\underline{n}$ from numerous observations made during seismic surveys [1, 9] is near 2.

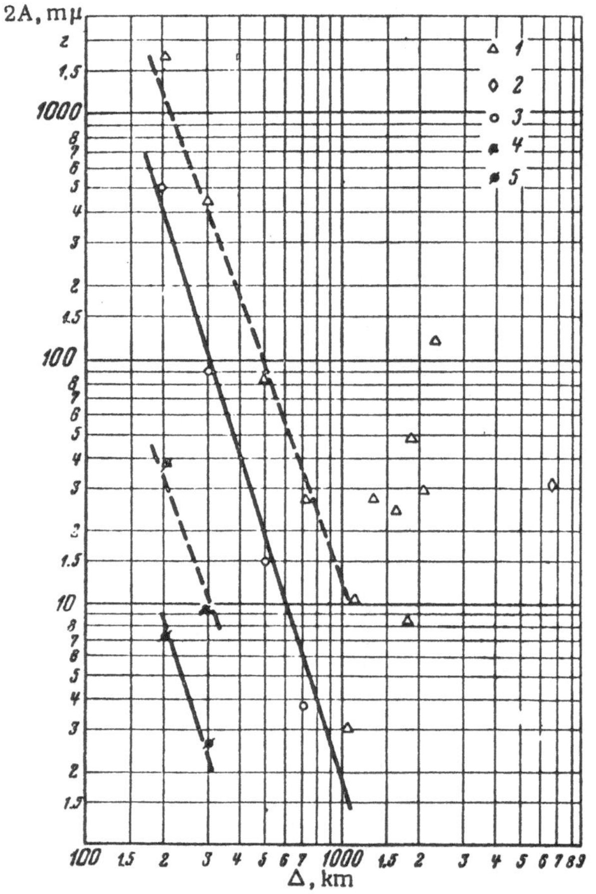

From the amplitude graphs of the Logan and Blanca shots and also of TNT explosions at Arys' and on the island of Helgoland, attempts were made to find values for the constants $\underline{n}$ and α in equation (1).

The values of $\underline{n}$ and α were determined by somewhat independent methods employed in a seismic survey [1, 7, 9].

These values were found to be the following for oscillations with periods of about 0.5-0.7 sec:

$$n \approx 2; \quad \alpha \approx 0.0025 \ \mathrm{km^{-1}}.$$

Equation (1), which expresses the character of the diminution in amplitude of the maximum phase of the Pn wave with distance for oscillations having periods of 0.5-0.7 sec, in this case takes the following form:

$$A_i \approx A_0 \left(\frac{\Delta_i}{\Delta_0} \right)^{-2} e^{-0.0025(\Delta_i - \Delta_0)}.$$

In order to make a more complete determination of the nature of change in the amplitude of the P wave with distance, we have plotted the data of all explosions on Fig. 25, including the experimental values of amplitude obtained at seismic stations in the USSR from large surface nuclear explosions detonated by the USA on the Marshall Islands in 1954. All these values were shifted to the level of the Logan shot by means of parallel transfer.

As seen from Fig. 25, the diminution in amplitude of the P wave is complex, and it is difficult to approximate a single unbroken curve for the entire interval of epicentral distances from 1200 to 10,000 km. The complex course of diminution in amplitude of the P wave with distance is probably due to the presence of zones in one of the earth's layers in which there is a sharp change either in the velocity or in the velocity gradient of seismic waves.

Fig. 24. Relationship between amplitudes of Pn and P waves and epicentral distances for the Logan and Tamalpais shots. 1) For the maximum peak of the Logan shot; 2) for the maximum peak of the Logan shot on records of the SVK-M seismograph; 3) for the first arrival of the Logan shot; 4) for the maximum peak of the Tamalpais shot; 5) for the first arrival of the Tamalpais shot.

It may be seen from an examination of the relationship between amplitudes of the Pn and P waves and distance, as shown in Figs. 23-25, that, after wave Pn gives way to wave P, approximately at a distance of 1200 km from the source, there is a slight increase in the amplitude of P as compared with Pn, but in comparison with values at greater distances, the amplitudes here are small, and a considerable variation in values is observed. At distances of about 2500 km the amplitude of the P wave reaches a maximum, and beyond that it decreases with distance.

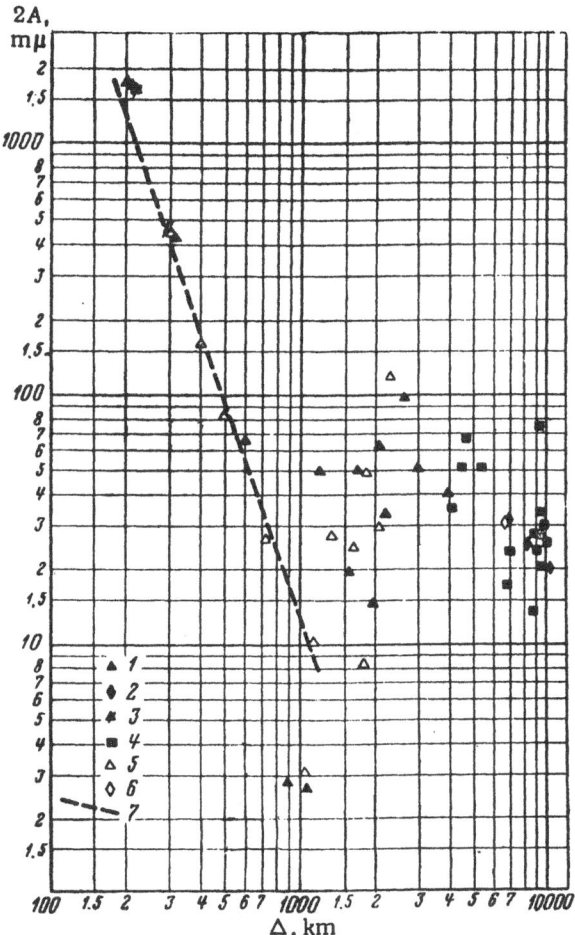

Fig. 25. Relationship between maximum amplitudes of Pn and P waves and the distance from explosions. 1) Blanca (Benioff seismograph); 2) Blanca (SVK-M seismograph); 3) Tamalpais (Benioff seismograph); 4) 1954 explosions on the Marshall Islands (SVK seismograph); 5) Logan (Benioff seismograph); 6) Logan (SVK-M seismograph); 7) averaging line.

Note. It should be noted that, in [34], double values of Pn amplitude are plotted on the curve that shows changes in the amplitude of Pn, similar to the curve of Fig. 23, whereas normal values of P are plotted on the curve for that wave.

9. Magnitude of Underground Nuclear Explosions

By analogy with earthquakes, the seismic effect of nuclear explosions are assigned magnitudes by American seismologists. No scale of magnitude has been yet developed for explosions; American seismologists therefore use scales developed by Gutenberg and Richter for earthquakes in evaluating the magnitude of underground nuclear explosions.

At epicentral distances up to 1000 km the M_L scale, proposed in paper [33], is used; this is based on the maximum amplitude on the record of horizontal Wood-Anderson torsional seismographs.

Beginning at epicentral distances of 1700 km one may use a scale discussed in the papers [21 and 25] for records of vertical and horizontal components of body waves according to the ratio of amplitude to period for the corresponding wave. Here m is the unifying value associated with M_S, determined by the amplitude-period ratio in surface waves of shallow earthquakes by the relationship indicated in [26]:

$$m = 0.63 M_S + 2.5; \tag{2}$$

whence $M_S = 1.27 (M_L-1) - 0.016 M_L^2$, or

$$m = 1.7 + 0.8 M_L - 0.01 M_L^2. \tag{3}$$

The values of M_L and m for the Blanca, Logan, and Rainier underground nuclear explosions were determined by American seismologists from records at a number of permanent and temporary seismic stations in the USA; the values are given in the paper [34]. They are reproduced here also in columns 4, 7, and 10 of Table 9.

The results of our determinations from seismograms sent us by the American delegation and from seismograms of Soviet stations are shown in columns 6, 9, and 12 of Table 9. The differences between the values are no greater than 0.2. Only the vertical component was used in determining m for the P wave.

The values of M_L were recomputed to m by equation (3); these are designated by m' in Table 9 (columns 5, 8 and 11).

In analyzing the values of m (for the Blanca and Logan shots) as shown in Table 9, one's attention is drawn to the following facts. The values determined from observations at stations at epicentral distances greater than 2500 km agree well among themselves and are greater than the values obtained at stations in the distance interval of 1700 to 2500 km. It is advisable to exclude from our consideration those values of m obtained at distances of 1700 to 2500 km where peculiarities may be observed in the nature of the changes in amplitude of P with distance. From the amplitude-distance relationship it is seen (Fig. 25) that in the 1200-2500 km interval there is actually a decrease in observed amplitude of the P wave in comparison with greater epicentral distances, and there is considerable variation among these values. American seismologists have also repeatedly pointed out the presence of distinguishing characteristics in the so-called shadow zone.

TABLE 9. Values of M_L and m for the Blanca, Logan, and Rainier Shots

No.	Station	Δ,km	Blanca from [34]		Blanca from measurements of m	Logan from [34]		Logan from measurements of m	Rainier from [34]		Rainier from measurements of m
			M_L	m'		M_L	m'		M_L	m'	
1	Tinemaha	181	5.1	5.5		4.8	5.3		4.2	4.9	
2	Woody	289	4.2	4.9		3.9	4.7		3.6*	4.4	
3	Riverside	371	4.7	5.2		4.4	5.0		3.8*	4.6	
4	Pasadena	382	4.8	5.3		4.4	5.0		4.0	4.7	
5	Mt. Hamilton	483	5.4	5.7	5.7	5.0	5.4	5.4	4.7	5.2	5.1
6	Barrett	503	3.9	4.7		3.7	4.5		3.3*	4.2	
7	Palo Alto	530	5.2	5.6	5.6	4.8	5.3	5.3	4.4	5.0	
8	Berkeley	540	4.9	5.4	5.2	4.5	5.1	5.0	4.1	4.8	
9	San Francisco	557	4.9	5.4		4.5	5.1		4.2	4.9	
10	Mineral	583	4.8	5.3	5.3	4.4	5.0	5.0	4.3	5.0	5.2
					m						
11	Tm 8—1700	1707	4.5		4.6						
12	Tm 8—1800	1804				3.8		3.6			
13	Dallas	1842	4.7								
14	Tm 9—1900	1902				4.1		4.3			
15	Tm 9—2000	2011	4.1		4.1						
16	Tm 10—2100	2111	4.5		4.7	4.0		4.1			
17	Tm 11—2200	2209	4.7		4.7						
18	Tm 22—2300	2305				4.9		4.7			
19	Tm 12—2500	2506				4.5					
20	Tm 12—2650	2665	5.1		5.3						
21	Tm 13—3000	3017	5.0		5.2						
22	Tm 15—4000	4021	5.1		5.3	4.9		5.0			
23	Tiksi	6890			5.2			4.9			
24	Temporary station	8300			5.1						
25	Temporary station	10 080			5.2						

* These data were presented by the American delegation in December, 1959.

Observations at epicentral distances up to 600 km are in good agreement with observations beyond the shadow zone. The Barrett and Woody stations are exceptions; these give a distinctly low value for all the explosions. A station correction should be introduced for the average magnitude values obtained from observations at these two stations for the explosions, or the data should be eliminated from consideration.

A detailed analysis of the facts noted above has been furnished in the paper of Yu. V. Riznichenko [10]. The following average magnitude values were obtained from that work:

Blanca	m = 5.2 ± 0.1
Logan	m = 5.0 ± 0.1
Rainier	m = 4.7 ± 0.1

where ± 0.1 is the standard deviation from the average.

In further discussions in this paper we shall use these values of m.

10. The Part of the Energy Going into the Formation of Seismic Waves during Underground TNT and Nuclear Explosions

There is considerable interest in determining the seismic energy given off during underground nuclear explosions, and also to discover what part of the energy goes into the formation of seismic waves during underground TNT and nuclear explosions. For this purpose a relationship is used that was established in seismology to relate the magnitude $\underline{m}$ with the energy of an earthquake E_s [26]:

$$\lg E_c = 5.8 + 2.4\,m. \tag{4}$$

In using this relationship we should keep in mind that, since (4) was established for earthquakes, we obtain only some relative value of the part of energy given off in the form of seismic waves when it is applied to explosions. At the same time, equation (4) obviously permits us to compare the ratios of seismic energy given off during TNT and nuclear explosions.

Below we site data on the magnitude $\underline{m}$ of two powerful underground blasts of chemical explosives, set off in recent years; we also determine the seismic energy given off during underground nuclear and TNT explosions.

TABLE 10. Amplitudes and Periods of Seismic Waves Recorded from the Arys' Blast, December 12, 1957, at 0.9 : 00 : 00

No.	Station	Δ, km	P_n, P^* T, sec	Surface wave T, sec	A_H, mμ	M_S	Type of seismograph
1	Namangan	255		2.0	5.33?	(3.8)	SK
2	Fergana	305		1.6	5.62	3.9	»
3	Andizhan	320		2.0	9.55?	(4.2)	»
4	Samarkand	330	0.7	2.0	5.95	4.0	»
5	Garm	370	0.7	1.9	4.26	3.9	»
6	Frunze	460	2.0?	2.25	2.14	3.7	»
7	Khorog	565	0.9	1.0?	2.35	3.81	»
8	Rybach'e	587	2.0?	2.0	4.9	4.1	»
9	Murgab	592		2.2	3.34	4.0	»
10	Alma-Ata	655	1.0	1.8	3.55	4.1	»
11	Temporary station	1140		2.0	0.71	3.7	»
12	Moscow	2600		2.5			»

Average value $M_S = 3.9 \pm 0.1$.

Explosion at Arys'. An experimental detonation of 1000 tons of ordinary chemical explosive was effected on December 19, 1957 at 09 : 00 : 00 Greenwich time in the Kabulsai district, along the Arys' segment of the Tashkent railroad (φ = 42°12'15.1" N, λ = 69°03'02.59" E). The principal explosive was ammonite No. 6.

The charge was placed in a special chamber in clay 40 m below the ground surface. The explosion produced a crater 200 m in diameter and about 30 m deep. The purpose of the detonation was scientific. Information on the seismic waves recorded from this explosion is contained in the paper [2]. Records at stations in Central Asia were made on D. P. Kirnos seismographs (SK).

The value of M_S on the scale of S. L. Solov'ev and N. V. Shebalin [12] was found from the maximum displacement of the surface wave without consideration of the period of the wave, since the calibration curve was obtained for periods characteristic of surface waves from earthquakes; for explosions the periods of these waves at the same epicentral distances are much smaller. The average value is $M_S = 3.9 \pm 0.1$. By substituting 3.9 for M_S in equation (2) we obtain m = 5.0.

The initial data on amplitudes and periods necessary for finding the magnitudes as well as the magnitude values thus obtained are given in Table 10.

Explosion at Pokrovsk-Ural'skii. A charge of chemical explosive (ammonite) weighing 3100 tons was detonated on March 25, 1958 at 09: 00: 00 at Pokrovsk-Ural'skii (φ = 60.2° N, λ = 59.9° E). The blast was part of an operation to excavate a canal about 1100 m long. The explosive charges were placed in 31 blast holes in rocky water-saturated soil at depths on the order of 20 m.

The longitudinal, transverse, and surface waves of the explosion were recorded. The longitudinal wave P was recorded at the greatest distance from the blast site, about 9000 km away. Data on arrival times, distances, and amplitudes of the seismic waves arising during this explosion are given in Table 11.

TABLE 11. Arrival Times, Amplitudes, and Periods of Seismic Waves Recorded from the Pokrovsk-Ural'skii Blast, March 25, 1957, 09: 00: 00

No.	Station	Δ deg	Δ km	Arrival time h m s		A, mμ	T sec	Type of seismograph	m
1	Sverdlovsk	4.3	480	P	09 01 08		2	Galitzin	
				S	01 53		2		
2	Moscow	12.5	1380	P	09 03 01		1	"	
				L	07 35		2		
3	Temporary station	13.0	1440	+ P	09 03 02.1	0.4	0.4	SVK-M	
				+ e	03 03.1	0.4	0.6	SGK-M	
				+ i	03 04.0	0.7	2.2		
				iS	05 20.6		1.2		
				L	06 34	~0.4	2.0		5.1
4	Semipalatinsk	13.5	1500	L	09 08 38	~0.1	1.5	SGK	4.9
5	Apatity	13.5	1500	+ P	09 03 15			SVK	
				e	05 41				
6	Pulkovo	14.6	1620	P	09 03 30			Benioff	
					06 02				
					06 50				
				L	07 47				
7	Sodank'yala	16.1	1790	— iP	09 03 46				
8	Helsinki	17.1	1900	iP	09 03 58				
9	Kiruna	18.9	2100	iP	09 04 17	0.1	1.5	Benioff	4.7
				e	04 21				
				iS	07 35				
10	Frunze	19.8	2200	L	09 10 20	0.07	1.5	SGK	5.0
11	Uppsala	20.7	2300	P	09 04 44	0.1	0.5	Benioff	5.4
				iS	08 26			SVK	
12	Simferopol	21.8	2420	P	09 04 55				
13	Skalstugan	22.5	2500	P	09 05 00				
				e	09 06				
14	Shilong	40.5	4500	— iP	09 07 47				
15	Tamanrasset	52.7	5860	eP	09 09 20				
16	Hungry Horse	72.7	8080	eP	09 11 27			Benioff	
17	Eureka	80.8	8980	eP	09 12 18			Benioff	
				e	12 25				

Average value $m = 5.0 \pm 0.2$

The periods of the longitudinal waves recorded at stations 1300 to 2300 km from the source were 0.5-1.5 sec. The periods of the surface waves recorded by the wide-band SVK and SGK instruments at distances of 2300 km were approximately 1.5-2.0 sec. The value of m obtained from data of the Kiruna and Uppsala stations is 5.1; although this value was obtained from the shadow zone we have used it because data are meager. The Semipalatinsk and Frunze stations, and also the temporary station, give $M_s = 4.0 \pm 0.1$. The average of all the data gives m = 5.0 $\pm$ 0.2.

The values of seismic energy E_s evolved from the three underground nuclear explosions and the two TNT explosions are given in the seventh column of Table 12. These values of E_s were determined by equation (4). In column 3-5 of Table 12 are given the TNT equivalents, the type of explosive, and the magnitude $\underline{m}$ for the indicated explosions; the eighth column supplies values of the ratio E_s/E_{tot} in %; and the ninth column indicates the maximum distance at which the P wave was recorded.

As may be seen from an examination of Table 12, approximately one percent of the total energy of the explosive goes into the formation of seismic waves when TNT is detonated underground. The proportion of total energy going into the formation of seismic waves when nuclear devices are detonated underground is approximately one-half to one-third this value, computed from the values of TNT equivalents to be approximately 0.2-0.3% of the total energy. The cited values of seismic energy of nuclear and TNT explosions are approximate, because of the small number of experimental data; no consideration at all has been given to the effect of the conditions of the ground at the explosion site. Furthermore, it should be noted that the chemical explosives discussed here, in contrast to the nuclear devices, were detonated in weak surface material; this might lower their seismic effect somewhat.

TABLE 12. Seismic Energy E_s Evolved from Underground Nuclear and TNT Explosions, the Energy of TNT Equivalent E_{tot}, and the E_s/E_{tot} Ratio for Three Nuclear and Two TNT Explosions

No.	Name of explosion	TNT equiv 10^3 tons	Type of explosion	m	E_{tot}, ergs	E_s, ergs	E_s/E_{tot}, %	Max. distance P wave recorded, km on seismograph	
								Benioff	SVK-M
1	Rainier	1.7	Nuclear blast	4.7	$7.2 \cdot 10^{19}$	$1 \cdot 10^{17}$	0.2	3700	
2	Logan	5	"	5.0	$2.1 \cdot 10^{20}$	$6 \cdot 10^{17}$	0.3		8300
3	Blanca	19	"	5.2	$8.1 \cdot 10^{20}$	$2 \cdot 10^{18}$	0.2		16 300
4	Arys'	1	Chemical explosive (ammonite No. 6)	5.0	$4.2 \cdot 10^{19}$	$6 \cdot 10^{17}$	1.4		2600
5	Pokrovsk-Ural'skii	3.1		5.0	$1.3 \cdot 10^{20}$	$6 \cdot 10^{17}$	0.5	9000	

11. Precision of Determining Coordinates of Epicenter

There is considerable interest in evaluating the precision with which the coordinates of epicenters may be determined by methods widely used in seismology.

The zero times at the epicenter and the coordinates of the epicenter were determined for the Logan shot. The arrival times of the Pn and P waves furnish the most reliable data for determining these values. It is difficult to distinguish the arrivals of transverse waves with sufficient precision at distances greater than 500 km.

There is a method, proposed by E. F. Savarenskii [11], for determining the epicenter of an earthquake by the absolute arrival times, t_1, t_2, and t_3 of the P wave at three seismograph stations. By assigning a time to the instant of the earthquake, t_0, and computing the differences $t_1 - t_0$, $t_2 - t_0$, and $t_3 - t_0$, it is possible to use the method of intersection to find from two pairs of stations two possible positions of the epicenter for the given t_0. By changing the value of t_0 we obtain two curves showing possible positions of the epicenter. The value of t_0 where these curves intersect gives the zero time at the epicenter, and the point of the intersection gives the coordinates of the epicenter. The use of a Wulff stereographic net for the necessary graphic construction permits this method to be used with little difficulty.

Any other wave on the seismogram, apart from P, may be used to determine t_0, although it will be very approximate.

In determining the coordinates of the epicenter of the Logan shot we proceeded in the following manner. The epicentral distances Δ_{L-P} for Hungry Horse and Fayetteville were determined very roughly from the differences in arrival times of the P waves and the surface waves L, and the corresponding times t_0 at the source were also computed

TABLE 13

Station	$\Delta°\,L-P$	t_0 h m s	$\Delta°$						
			59 m 55 s	60 m 05 s	60 m 15 s	60 m 25 s	60 m 35 s	60 m 45 s	60 m 55 s
Hungry Horse	10	06 00 17	11.6	10.9	10.2	9.4	8.7	8.0	7.2
Fayetteville	15	06 00 34	18.1	17.3	16.5	15.7	14.9	14.2	13.4
College			34.0	32.8	31.6	30.5	29.4	28.3	27.2

(Table 13). These data from the two stations give $t_0 = 06:00:25$; the probable accuracy of this determination is ± 30 sec. By changing the value of t_0 10 sec we obtained, by the above described method of intersecting lines, an epicenter with the coordinates 36.5° N and 116° W for a $t_0 = 05:59:55$.

All determinations of Δ and t_0 from the Pn and P waves were made with the Jeffreys-Bullen travel-time curves for a surface focus.

The deviation δt_p in the experimental values of travel time for the P wave (with a correction for ellipticity) from these travel-time curves averages less than a second; this deviation is probably related to the local geologic conditions at the stations and to errors in observation, since the deviation varies in sign (Table 14). It would be, of course, better to use a regional travel-time curve for the Pn wave. For most stations the observed travel time of the Pn wave averaged approximately a second less than indicated on the mean travel-time curve. This deviation does not exceed the limits of accuracy in our observations, and we apparently introduced no large error in using the Jeffreys-Bullen travel-time curve for the Pn wave. Besides, in this report we think it best to examine the least favorable possibility when no regional travel-time curve is known.

The epicenter was determined for the time $t_0 = 05:59:55$ from data from stations for which the arrival times of the P wave have been listed in Tables 2 and 5; the computations were made by the intersection method on a Wulff net. The coordinates of this epicenter were found to be $\varphi_e = 37.0°$ N and $\lambda_e = 116.5°$ W.

Epicentral distances computed by the equation

$$\cos \overline{\Delta} = \sin \varphi_e \cdot \sin \varphi + \cos \varphi_e \cdot \cos \varphi \cdot \cos (\overline{\lambda}_e - \lambda),$$

where φ and λ are the coordinates of the station and $\overline{\varphi}_e$ and $\overline{\lambda}_e$ are the coordinates of the epicenter, permitted us to define the time for this epicenter more precisely. From the data of 30 stations located more than 300 km from the source, the time was found to be $\overline{t}_0 = 06:00:00 ± 02$ sec.

We did not consider data from stations less than 300 km from the source, since the method of successive approximations [11] used in further computation is valid only for epicentral distances much greater than the distance between actual and approximate epicenters; in our work this distance proved to be 30 km.

The system of equations in the form

$$x_0 \sin Az + y_0 \cos Az = \alpha,$$

was solved by the method of least squares, relative to x_0 and y_0, where Az is the azimuth from the epicenter to the station, $\alpha = \Delta_0 - \overline{\Delta}$, Δ_0 is the epicentral distance from the travel-time curve obtained from the recorded arrival times of Pn and P at the station and from the assumed time $\overline{t}_0$, $\overline{\Delta}$ is the computed epicentral distance, and x_0 and y_0 are corrections permitting one to define the position of the epicenter more precisely:

$$\varphi_e = \overline{\varphi}_e - y_0; \quad \lambda_e = \overline{\lambda}_e - \frac{x_0}{\cos \varphi_e}.$$

The computations from 30 stations (Table 14) gave for the corrections the values $y_0 = -0.09°$ and $x_0 = 0.22°$. Thus, the more precise coordinates for the epicenter of the Logan shot are $\varphi_e = 37°05'$ N and $\lambda_e = 116°13'$ W. The error in locating the epicenter amounts to about 10 km. The time of detonation at the epicenter was $t_0 = 06:00:00 ± 01$ sec, 26 of the 30 stations agreeing with this value within ± 2 sec.

TABLE 14

No.	Station	$\delta t_p = t_{exp} - t_{graph}$, sec	Az°	37°,0N; 116°,5 W; t_o = 06 00 00			37°05' N; 116°13' W	
				$\overline{\Delta}°$	$\Delta_0°$	$\alpha°$	$\overline{\Delta}°$	h m s (t_o)
1	Fresno	0	260	2.6	2.9	0.3	2.9	06 00 00
2	Fort Tejon	0	225	2.9	3.2	0.3	3.1	01
3	Dalton	—1	205	3.0	3.2	0.2	3.2	01
4	Mt. Wilson	—1	210	3.0	3.2	0.2	3.2	01
5	Riverside	—1	200	3.1	3.2	0.1	3.2	01
6	Pasadena	0	210	3.2	3.3	0.1	3.3	01
7	Hayfield	—1	175	3.4	3.4	0.0	3.4	01
8	Reno	—1	310	3.6	3.6	0.0	3.7	05 59 59
9	Palos Verdes	+1	210	3.6	3.9	0.3	3.8	06 00 01
10	Palomar	—2	190	3.6	3.8	0.2	3.8	00
11	Mt. Hamilton	—2	275	4.1	4.3	0.2	4.3	00
12	Barrett	—1	185	4.3	4.4	0.1	4.4	00
13	Berkeley	—1	280	4.6	4.8	0.2	4.8	01
14	Mineral	—1	310	5.2	5.2	0.0	5.3	05 59 59
15	Tucson	—2	135	6.6	6.5	—0.1	6.6	58
16	Boulder	0	70	9.3	9.0	—0.3	9.0	06 00 00
17	Laramie	—1	60	9.5	9.1	—0.4	9,2	05 59 59
18	Butte	+2	15	9.5	9.4	—0.1	9,3	06 00 02
19	Boseman	+3	25	9.6	9.6	0.0	9.4	02
20	Hungry Horse	—1	10	11.5	11.2	—0.3	11.4	05 59 58
21	Rapid City	—2	50	12.3	11.8	—0.5	12.1	56
22	Lubbock	+3	115	12.4	12.4	0.0	12.2	06 00 03
23	Fayetteville	0	85	17.9	17.7	—0.2	17.7	00
24	Florissant	+1	80	20.6	20.5	—0.1	20.4	01
25	St. Louis	0	80	20.7	20.5	—0.2	20.5	00
26	Ottawa	0	65	31.4	31.3	—0.1	31.2	01
27	College	0	335	33.4	33.4	0.0	33.4	00
28	Seven Falls	—3	60	34.8	35.0	+0.2	34.6	04
29	Tiksi	—2.6	340	62.2	61.9	—0.3	62.2	05 59 57
30	Temporary station	—0.9	320	74.9	74.9	0.0	74.8	06 00 00

Similar determinations for the Blanca shot give the same results, since the observations for both shots were made at the same stations and with identical precision.

Thus, the position of the epicenter of the explosion according to data taken from stations surrounding the shot, and using the mean travel-times on the Jeffreys-Bullen graph for foci of zero depth may be placed inside a circle the radius of which is approximately 10 km, i.e., within an area of about 300 km^2; in these computations the accuracy of zero time is ± 1 sec. If a regional travel-time curve were used the precision of locating the epicenter should be improved.

Conclusions

Our treatment of the seismic data of the Rainier, Tamalpais, Logan, and Blanca underground nuclear explosions and of the Arys' and Pokrovsk-Ural'skii underground blasts of chemical explosives permits us to make the following conclusions.

1. Shock waves produced by underground nuclear and chemical explosions are detected at rather great distances from the site of detonation. The approximate distances at which explosions of various yields may be detected, so far as now know, are the following:

1) for nuclear explosions:

0.5 kilotons	300 km (Benioff seismograph)	
1.7 "	3700 " (" ")	
5 "	8300 " (SVK-M ")	
19 "	16,000 " (" ")	

2) for chemical explosions:

1 kiloton	2600 km (SVK-M seismograph)
3 "	9000 " (Benioff ")

2. The Pn wave is recorded among the first arrivals at epicentral distances of 200-1100 km, the Pat distances of 1200-10,000 km, and the PKP at distances greater than 16,000 km.

The transverse waves S and S* from nuclear explosions are distinguished on records of the Benioff seismograph at distances of 200-500 km. The direct transverse wave S from the Pokrovsk-Ural'skii chemical explosion (3 kiloton) was recorded on a Benioff seismograph at a distance of 2300 km.

Surface waves were recorded at distances up to 2000-3000 km.

3. The direction of movement corresponding to the compressional phase was recorded in the first arrival of the longitudinal wave at distances up to approximately 700 km for the Logan shot and up to 1000 km for the Blanca shot, on Benioff seismographs at stations in the special network. The amplitude corresponding to this compressional phase was not great.

By correlations of the initial parts of the records it has been ascertained that the absence of the compressional phase at large epicentral distances is due to loss of the first peak on the records. This loss of the first peak is probably the result of rapid damping of the high-frequency oscillations characteristic of underground explosions, and also to the distorting effect of the recording apparatus.

In order to distinguish explosions from earthquakes by the first arrivals it is suitable to use the method adopted for studying the mechanism of earthquakes. In contrast to ordinary earthquakes, the distribution of signs (directions of movement) for the Logan and Blanca shots do not permit one to construct nodal lines.

4. The periods of the longitudinal Pn and P* waves ranged from 0.5 to 0.8-1.0 sec at distances of 200 to 1000 km on records of nuclear explosions made by Benioff seismographs; the range was from 0.2 to 0.8 sec for chemical explosions as recorded on a D. P. Kirnos seismograph, and from 0.6 to 2.5 sec for earthquakes of approximately the same level as recorded on this latter instrument.

The periods of the transverse Sn and S* waves ranged from 0.6 to 1.2 sec for nuclear explosions on the records of Benioff seismographs at distances of 200-500 km. For earthquakes of approximately the same energy level, the periods ranged from 1 to 4 sec on the records of a D. P. Kirnos seismograph at distances of 100-1000 km.

Consequently, the periods of body waves from underground explosions are somewhat smaller than from earthquakes of the same energy level.

5. The periods of surface waves from nuclear explosions, recorded on Benioff seismographs, are the same as for longitudinal waves at epicentral distances of 100-500 km (0.7-1.2 sec); at distances of 500 to 2000 km periods of 2-3 sec were noted on records of seismographs having a wider range than Benioff seismographs. Long-period seismographs at Pasadena, Palisades, and Berkeley recorded surface waves with periods of 7-10 sec.

A comparison of the periods of surface waves recorded on a D. P. Kirnos seismograph for chemical explosions and for earthquakes of the same energy level shows that the two are substantially different. The periods of surface waves from chemical explosions are 2.0 ± 0.5 sec, and are practically invariant with distance. At a distance of about 1000 km this value is but one-fourth the period of a surface wave from an earthquake.

The period of a surface wave may serve as one of the criteria for distinguishing the record of an explosion from the record of an earthquake.

6. The character of the changes in amplitude with distance varies for different waves.

The amplitude of the Pn wave for nuclear explosions with a period of 0.5-0.7 sec decreases with distance according to the following law:

$$A_i = A_0 \left(\frac{\Delta i}{\Delta_0} \right)^{-2} e^{-0.0025(\Delta_i - \Delta_0)}.$$

The character of the changes in amplitude of the P wave is more complex. At distances of 1200-2500 km the amplitude is less than at greater distances; a marked variation in values has been observed in the indicated interval of epicentral distances. The maximum amplitude of the P wave is recorded at a distance of about 2500 km; beyond this distance the value gradually decreases.

7. The nature of the seismic records, the types of waves recorded, the predominant periods, and other characteristics are practically identical for nuclear and chemical underground explosions. The seismic effect of chemical explosions, without considering the different conditions of the ground at the detonation site, is approximately two to four times the effect of nuclear explosions.

An experimental check on the effectiveness of the control system may be made by detonations of ordinary explosives.

8. The location of the epicenters of the Logan and Blanca underground explosions, from data of seismic stations surrounding the epicenter and from the Jeffreys-Bullen mean travel-time curve (i.e., with no known regional travel-time curve), was determined within an area of about 300 km^2.

LITERATURE CITED

1. I. S. Berzon, "Determination of the exponent of the scattering function for refracted waves from experimental data," Izv. AN SSSR, seriya geofiz., No. 4 (1951).

2. E. M. Butovskaya, V. I. Ulomov, M. A. Dzhunisov, and V. N. Yakovlev, Travel-Time Curve for Seismic Waves and Some Peculiarities in Structure of the Earth's Crust in Central Asia according to Data from the Records of High-yield Explosions (A Report) [in Russian] (Institut matematiki AN Uzb. SSR, 1958).

3. O. D. Gotsadze, V. I. Keilis-Borok, I. V. Kirillova, S. D. Kogan, T. I. Kukhtikova, L. N. Malinovskaya, and A. A. Sorskii, Investigations of the Mechanism of Earthquakes [in Russian] (Tr. Geofiz. inst., No. 40 (166) (1957).

4. "Report of a committee of experts on the study of methods of detecting infractions of possible agreements to stop nuclear testing," Atomnaya energia 5, No. 4 (1958).

5. V. I. Keilis-Borok, "Differences in the spectra of surface waves from earthquakes and from underground explosions," Present volume.

6. S. D. Kogan, I. P. Pasechnik, and D. D. Sultanov, "Differences in periods of seismic waves produced by underground explosions and by earthquakes," Doklady Akad. Nauk SSSR 129, No. 6, (1959).

7. T. I. Oblogina, "A new method of determining the coefficient of seismic waves," Prikladnaya geofizika, No. 27 (1960).

8. I. P. Pasechnik and N. E. Fedoseenko, "An experiment to modernize SVK and SGK seismographs," Izv. AN SSSR, seriya geofiz., No. 12 (1959).

9. Yu. V. Riznichenko, Scattering and Absorption of Seismic Waves [in Russian], Tr. Geofiz. inst., No. 35 (162) (1956).

10. Yu. V. Riznichenko, "The seismic magnitudes of underground nuclear explosions." Present volume.

11. E. F. Savarenskii and D. P. Kirnos, Elements of Seismology and Seismometry [in Russian], Gostekhizdat, (Moscow-Leningrad, 1949).

12. S. L. Solov'ev and N. V. Shebalin, "Determination of earthquake intensities by displacement of ground in surface waves," Izv. AN SSSR, seriya geofiz., No. 7 (1957).

13. F. I. Monakhov, I. P. Pasechnik, and N. V. Shebalin, Seismic Stations of the USSR Working on the IGY Program [in Russian], Izd. AN SSSR (1959).

14. AEC Releases. Data on Hardtack Bomb Tests, Tuesday, March 10, 1959.

15. L. F. Bailey and K. F. Romney, "Seismic waves from the Nevada underground explosion of September 19, 1957," Bull. Geol. Soc. Am. 69, No. 12, pt. 2 (1958).

16. P. S. Carder and L. E. Bailey, "Seismic wave travel times from nuclear explosions," Bull. Seism. Soc. Amer. 48, No. 4 (1958).

17. Control and Reduction of Armaments: Detection of and Inspection for Underground Nuclear Explosions. Staff Study. No. 10, Subcommittee on Foreign Relations, U. S. Senate, June 23, 1958.

18. Disarmament and Foreign Policy Hearing before the Subcommittee on Foreign Relations, U. S. Senate 96th Congress, 1st Session, Pt. 1, January 28, 30, and February 2, 1959. Washington, D. C. (U. S. Government Printing Office, Washington, 1959).

19. D. S. Carder and W. K. Cloud, "Surface motion from large underground explosions," Journ. Geophys. Res. 64, No. 10 (1959).

20. B. F. Grossling, Seismic Waves from Underground Atomic Explosion in Nevada, "Bull. Seism. Soc. Amer.," 49, January, 1959, p. 11-32.

21. B. Gutenberg, Amplitudes of P; PP and Magnitude of Shallow Earthquakes, "Bull. Seism. Soc. Amer.," 35, No. 2 (1945).

22. B. Gutenberg, Seismic Waves from the Atomic Bomb Tests, "Trans. Amer. Geophys. Union," 27, No. 6 (1946).

23. B. Gutenberg, Travel Times from Blasts in Southern California, "Bull. Seism. Soc. Amer.," 41, No. 1 (1951).

24. B. Gutenberg, Wave Velocities in the Earth Crust, "Geol. Soc. Amer.," Spec. Paper 62, No. 19634 (1955).

25. B. Gutenberg and C. F. Richter, Earthquake Magnitude, Intensity, Energy and Acceleration (Second Paper), "Bull. Seism. Soc. Amer.," 46, No. 2 (1956).

26. B. Gutenberg, Magnitude and Energy of Earthquakes, "Ann. Geofis.," Rome 9, No. 1 (1956).

27. H. Jeffreys and K. E. Bullen, Seismological Tables (London, 1940).

28. G. W. Johnson, G. T. Pelsor, R. G. Preston, and C. E. Violet, The Underground Nuclear Detonation of September 19, 1957; Rainier, Operation Plumbbob. Univ. of California, Radiation Laboratory, (UCRZ 5124), February 4, 1948.

29. G. W. Johnson and C. E. Violet, Phenomenology of Contained Nuclear Explosions, Univ. of California, Lawrence Radiation Laboratory, (UCRZ 5124), December, 1958 .

30. G. C. Kennedy and G. H. Higgins, Temperatures and Pressures Associated with the Cavity Produced by the Rainier Event, (UCRZ 5281), Chemistry General C-4, July, 1958 .

31. Nuclear Explosions in 1954, 1956, and 1957, "Bull. Seism. Soc. Amer.," 48, July, 1958, p. 283.

32. D. S. Carder, W. K. Cloud, T. H. Pearce, and L. M. Merphy, Surface Motions from a Series of Underground Nuclear Tests, ITR-1705 (1958).

33. C. F. Richter, An Instrumental Earthquake Magnitude Scale, "Bull. Seism. Soc. Amer.," 25 (1935).

34. K. F. Romney, Amplitudes of Seismic Body Waves from Underground Nuclear Explosions, "Journ. Geophys. Res.," 64, 10 (1959).

35. United States Atomic Commission Releases Data on Hardtack Bomb Tests, Earthquake Notes, Eastern Section, "Seism. Soc. Amer.," 29, 4 (1958).

SEISMIC MAGNITUDES OF UNDERGROUND NUCLEAR EXPLOSIONS

Yu. V. Riznichenko

This paper discusses the results of determining magnitudes of nuclear explosions set off at the test site in Nevada (USA) in 1957-58. The average magnitudes of these explosions have been computed. The relationship between yield of an explosion (in kilotons of TNT equivalent) and average magnitude of the explosion has been determined. The annual number of shallow-focus earthquakes exceeding in magnitude explosions of a given yield is given. It is shown that this number is no greater than the value adopted by the committee of experts for stopping nuclear tests that met at Geneva in 1958.

Introduction

The chief difficulty in detecting underground nuclear explosion by seismic methods—distinctive methods that permit their use at great distances—lies in the necessity of distinguishing the seismic effect of an explosion from that of a natural earthquake.

From seismology it is known that the annual number (frequency) of earthquakes for the world at large is greater for quakes of smaller magnitudes, i.e., for weaker earthquakes. The seismic effect of an explosion, for comparison with earthquakes, may also be given a tentative designation of magnitude. It is obvious that the lower the magnitude of the underground nuclear explosion the greater the number of earthquakes that it may be confused with, and, consequently, the more difficult it becomes to distinguish the nuclear blast.

In order to appraise the possible effectiveness of the proposed international control system, it is important to know the annual number of earthquakes throughout the world corresponding to, or, more precisely, exceeding in magnitude underground nuclear explosions of any given yield.

All these aspects of the problem of detecting underground nuclear explosions are discussed in the present paper.

1. Magnitudes of the Explosions

The magnitude of an explosion or of an earthquake is something of a generalized characteristic of the focus of the particular seismic phenomenon. Our concept of the magnitude of an earthquake was initially established by Richter in 1935 [1], and since that time it has been repeatedly re-examined and improved [2, 3, and others]. According to Richter [4], in the simplest case (for a local zone), the magnitude M is determined by the equation

$$M = \log A - \log A_0,$$

where A is the amplitude of the peak on the seismogram of a given earthquake, recorded at a given epicentral distance $\underline{x}$ by means of some standard instrument, and A_0 is the similar amplitude for some standard earthquake. Both values, A_0 and A, are reduced to the standard distance x = 100 km. This reduction is accomplished by means of an experimentally established relationship between the amplitude A and the epicentral distance. The zero level A_0 (at) x = 100 km) was tentatively adopted as 10^{-3} mm for maximum amplitudes on the record of a Wood-Anderson torsion seismograph. In keeping with this determination, the magnitude, at least in principle, should not depend on the distribution of observing stations. Later the magnitude was related to the full seismic energy at the earthquake focus, i.e., with the energy of the seismic waves emanating from the focal zone and traveling to great distances.

At present there are several magnitude scales for earthquakes, based on values of the amplitude A or of the ratio A/T, where T is the period. These scales are distinguished from each other by the types of waves used, by the ranges of their applicability, and, generally speaking, by the numerical results obtained. A definite correlation among the various magnitude scales has been established in recent years. A good summary of the present status of this problem is given in the recently published book of Richter [4].

The following three scales of magnitude are the most widely used in practice.

1. Local magnitudes M_L. These magnitudes are measured in the "local" zone, beginning near the epicenter and extending to distances of x = 1000-1100 km, to the edge of the well-known "shadow zone." In computing M_L the maximum amplitudes are used from records of standard Wood-Anderson torsional seismographs. These amplitudes correspond to "shear" waves—transverse or surface, which are not distinguished in the local zone near the epicenter.

2. The magnitudes M, or, as sometimes designated, M_s, determined by surface waves. These waves begin to appear as distinctive parts of the record in the local zone not too near the epicenter, are observed beyond in the "shadow zone" (zone of diffraction), and are also found in the more remote zone beyond. This has led Gutenberg [3] to use the M scale for tying the other two scales together: the magnitude M_L determined in the local zone up to the shadow zone and the teleseismic magnitude $\underline{m}$ determined in the zone beyond the shadow zone.

3. Teleseismic magnitudes $\underline{m}$. These are determined in the remote zone, which begins with x = 1700 km according to Gutenberg. For explosions, the "remote zone," as pointed out below, may be reasonably considered to be the zone beginning at a distance of approximately x = 2500 km. The beginning of the zone for which there already exists a scale of teleseismic magnitudes is referred properly to a continuation of the zone of diffraction, which extends from the beginning of the shadow zone (x = 1000-1100 km), includes the shadow zone, and embraces the zone beyond (where the seismic signal is received), extending to an epicentral distance approximately of x = 2500 km. In computing the teleseismic magnitude $\underline{m}$, the amplitudes and periods of body waves—longitudinal and transverse —are used. On the basis of this scale Gutenberg has developed a "unified magnitude scale" such that the teleseismic magnitudes, which we use in our further discussions, coincide with the unified values.

The unified magnitude scale of $\underline{m}$ is at present the most theoretical scale and is related to the principal physical value that characterizes the seismic effect at the focus: the seismic energy E. In addition, it is very important for comparing explosions and earthquakes, as discussed in this paper, to have the latest and most reliable statistics on the annual number of earthquakes in the earth, with their distribution according to magnitude, as plotted by Gutenberg [3] on the basis of the unified magnitudes $\underline{m}$.

The problem of total value of the seismic effect at an earthquake focus is now being actively investigated in seismology. As material from instrumental records accumulates, the magnitude scales and the relationships tying them together are made more precise. Some relationships are now known much less precisely than we would like them to be, and they will undoubtedly be refined in the future. Nevertheless, in the absence of anything better, we must use what we know at the present time. It is important that, when making computations in this rapidly improving field, we base our work on modern concepts and do not restrict ourselves to old techniques that may not be in agreement with later work of very fundamental importance. It would seem obvious that the essential virtue of this position in making our investigations is brought to light by what follows.

Let us note that there is a tendency in the USSR to change from making determinations of conditional values — magnitudes— to direct determination of the more distinct physical value— seismic energy at the focus, relative to a sphere of established radius surrounding the focus. In addition, it is important to consider the frequency spectrum of the focus [5, 6]. This has special significance in comparing the seismic effects of earthquakes and of explosions, which, as a rule, are sources of higher-frequency waves [19, 20]. The author is inclined to believe that the energy-spectrum trend is more gradual. In the present paper, for convenience in comparing our work with results of foreign investigators, we have kept, for the time being, the more widespread method of designating the value of an earthquake by its magnitude. From this value it is possible, when desirable, to approximate the seismic energy at the focus [4] and, further, to relate the value to the average frequency spectrum of the earthquake [6].

When the ordinary method of determining earthquake magnitudes was applied to underground explosions, a number of peculiarities were encountered, which at first sight appear paradoxical. These peculiarities have proved to be related principally to the fact that, when the magnitude is computed, the effects of the different frequency spectra of these sources on the seismic waves are not properly taken into account, nor is the relationship to these effects of the different damping characteristics of the seismic waves with distance from the epicenter, particularly in the zone of diffraction. Another factor influencing the results involves differences in distribution of the total seismic energy of an explosion or an earthquake among waves of different types.

Initial Data on the Magnitudes of Underground Nuclear Explosions. The first underground nuclear explosion, the Rainier shot, with a TNT equivalent of 1.7 kiloton, was set off at the test site in Nevada, USA, in September, 1957. It was part of the Plumbbob operation. The depth of the detonation chamber was about 300 m, and the surrounding rocks were bedded volcanic tuffs. The magnitude M_L of the Rainier shot was determined by American specialists to be 4.25 ± 0.4 [7]. This value was adopted as a basis for computations by the experts at Geneva in the summer of 1958. This first conference of experts, as is well known, was terminated by mutual agreement on all the basic problems, and its recommendations were approved by the governments of the USSR, the USA, England, and other countries participating in the conference.

Soon afterward, in October 1958, the Americans set off a series of five nuclear explosions at the same test site in Nevada; these were part of the Hardtack II series [8, 9]. This series included two larger underground explosions: the Blanca, with a TNT equivalent of 19 kilotons (initially evaluated at 23 kilotons), and the Logan, with an equivalent of 5 kilotons. These explosions, as the earlier Rainier shot, were set off at a depth of about 300 m. The remaining three shots were much smaller, each having a yield of less than 0.1 kiloton.

Of the three underground nuclear explosions thus far set off, only three—Rainier, Logan, and Blanca—belong to the range of yields, from 1 to 20 kilotons discussed by the experts in Geneva for the purpose of possible establishment of control measures. Still, one of the small explosions of the Hardtack II series, the Tamalpais shot (with an equivalent of 0.072 kiloton and a depth of detonation of about 100 m), may be used for some comparisons with the shots in the principal size range. The other two small shots proved to be unsuitable for this purpose. Actually, the Evans shot (0.055 kiloton, detonation depth of about 300 m), despite the slight difference in yield from the Tamalpais, for reasons not clear to the Americans themselves [9], produced a very small seismic effect: the amplitudes of the waves from it were less than one-tenth those from the Tamalpais shot. This phenomenon was given a very poetic name by the Americans: the Evans Mystery. The Neptune shot (0.090 kiloton), set off at a depth of only 30 m, caused extensive ejection into the atmosphere, and, for this reason, the shot, properly speaking, cannot be included among the underground explosions.

Apparently the explosions in the chief range of yields may be considered to have been rather well camouflaged seismically; that is, they were proper underground shots. It is true that one of these, specifically the strongest, the Blanca, was accompanied by some loss into the atmosphere: later, more than ten seconds after the instant of detonation, ejection of gas and dust was noted over the epicenter. It is assumed, however, that this ejection was due to the collapse of rock into the cavity formed by the blast, after the pressure in the cavity had begun to fall. The principal strong seismic shock, the waves of which were recorded at great distances, was associated with the very instant of the detonation. It is difficult to believe that the size of this shock might prove to be much less if the explosion were not completely camouflaged.

Data on the measured magnitudes of the large underground nuclear explosions of the Hardtack II series—Blanca and Logan— together with some "average" magnitude values for these explosions and a somewhat "revised" magnitude for the Rainier explosion were first published in the "Work Report" of January 5, 1959 [10]. This report presents the well-known treatment of the American "New Seismic Data," around which there later developed widespread political, and then technical, discussion. The same data on magnitudes were presented in a paper by Romney [9] and in a report by this author at the second conference of experts, which met in Geneva at the end of 1959.

A more complete summary of the data on measurements of magnitudes of these explosions was presented by the American specialists at this conference on December 14, 1959. We have reproduced the data from this summary for the Rainier, Logan, and Blanca shots in Table 1 (M_L values) and Table 2 ($\underline{m}$ values); these form the basis for all our subsequent computations involving these explosions.*

---

*From Tables 1 and 2 of the December 14 summary we exclude only one figure, specifically the individual value of local magnitude for the Blanca shot $M_L = 4.9$, obtained at the San Nicolas station 540.0 km from the epicenter. Only for this station was there no information on magnitudes of the two other shots, and this makes any effective use of this figure in the computations difficult. Inclusion of this value in the computation would change the average local magnitude of the Blanca explosion only 0.03, which is of no practical significance.

TABLE 1. Local M_L and the Unified $\underline{m}$ Magnitudes of the Blanca, Logan, and Rainier Shots from Observations in the First Zone

No.	Station	Distance km	Magnitude					
			Blanca, 19 km		Logan, 5 km		Rainier, 1.7 km	
			M_L	m	M_L	m	M_L	m
1	Tinemaha	180.7	5.1	5.5	4.8	5.3	4.2	4.9
2	Woody	289.1	4.2	4.9	3.9	4.7	3.6*	4.4
3	Riverside	370.8	4.7	5.2	4.4	5.0	3.8*	4.6
4	Pasadena	382.2	4.8	5.3	4.4	5.0	4.0	4.7
5	Mt. Hamilton	482.8	5.4	5.7	5.0	5.4	4.7	5.2
6	Barrett	502.8	3.9	4.7	3.7	4.5	3.3*	4.2
7	Palo Alto	530.4	5.2	5.6	4.8	5.3	4.4	5.0
8	Berkeley	540.5	4.9	5.4	4.5	5.1	4.1	4.8
9	San Francisco	556.6	4.9	5.4	4.5	5.1	4.2	4.9
10	Mineral	583.0	4.8	5.3	4.4	5.0	4.3	5.0
	Average magnitude		4.79	5.30	4.44	5.04	4.06	4.77
	σ_{M_L}, σ_m		±0.45	±0.31	±0.40	±0.28	±0.41	±0.30
	$\sigma_{\overline{M}_L}, \sigma_{\overline{m}}$		±0.14	±0.10	±0.13	±0.09	±0.13	±0.10

*These values were not used in the initial determination in 1958 of the magnitude M_L for the Rainier shot.

The local magnitudes M_L in Table 1 were computed by American specialists [19] from maximum amplitudes of "shear" waves in the first local zone within the epicentral distances of x = 180.7 and x = 583 km. Records of Wood-Anderson torsional seismographs were used, the same instruments for which the initial scale was worked out and for which the procedure of determining magnitudes was described by Richter as early as 1935 [1]. This procedure is still used for determining M_L [9].

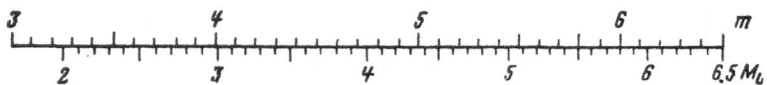

Fig. 1. Nomogram for recomputing the local magnitude M_L to the unified magnitude $\underline{m}$ (or vice versa) according to the Gutenberg formula $m = 1.7 + 0.8\,M_L - 0.01\,M_L^2$.

On the basis of these values of M_L we have computed the corresponding values of the magnitude $\underline{m}$ on the unified scale of Gutenberg. The change from one scale to the other was made by the well-known equation [3, 4]:

$$m = 1.7 + 0.8M_L - 0.01M_L^2. \tag{1}$$

Figure 1 shows the corresponding nomogram.

The conversion of M_L to $\underline{m}$ was made for all values, in order that all the magnitude values obtained in the local (Table 1) and the distant (Table 2) zones be expressed in a single system of units and in order that the explosions might be compared with Gutenberg's statistics on earthquakes [3].

The teleseismic magnitudes $\underline{m}$, shown in Table 2, were computed in the USA [19] from the amplitudes of the longitudinal wave P recorded at American stations on Benioff seismographs (zone 2, Nos. 1-8; zone 3, Nos. 1-4); the

TABLE 2. Teleseismic (Unified) Magnitudes $\underline{m}$ of the Blanca and Logan Shots from Observations in the Second and Third Zones

No.	Station	Distance, km	Magnitude Blanca, 19 km	Magnitude Logan, 5 km
		Zone 2		
1	Temporary	1706.7	4.5	
2	»	1803.7		3.8
3	»	1842.0	4.7	
4	»	1902.1		4.1
5	»	2011.2	4.1	
6	»	2111.3	4.5	4.0
7	»	2208.8	4.7	
8	»	2305.0		4.9
	Average magnitude		4.50	4.20
			$\sigma_m \pm 0.24$	± 0.48
			$\sigma_{\overline{m}} \pm 0.11$	± 0.24
		Zone 3		
1	Temporary	2506.0		4.5
2	»	2665.3	5.1	
3	»	3017.4	5.0	
4	»	4020.6	5.1	4.9
5	Tiksi	6890	5.2	4.9
6	Temporary	8320	5.1	
7	»	10080	5.2	
	Average magnitude		5.12	4.76
			$\sigma_m \pm 0.08$	± 0.23
			$\sigma_{\overline{m}} \pm 0.03$	± 0.13

computations were made by means of the procedure and tables described in [3]. Three later values of $\underline{m}$ were obtained at very distant Soviet stations (zone 3, Nos. 5-7). For these the record of the P wave was used, with due consideration to amplitudes and periods, as recorded on SVK-M seismographs.

All the magnitudes of $\underline{m}$ in Table 2 agree completely with the unified scale of Gutenberg and require no further recomputation. In these same units $\underline{m}$ has been compared with Gutenberg's earthquake statistics [3].

Consideration of the Data on Magnitudes. From the seismic energy E of the first underground nuclear explosion, the Rainier shot, determined in the local zone, Carder and Cloud computed the initial magnitude of this explosion[12], which proved to be

$$M_L = 4.6 - 4.7.$$

In this computation they used the squared relationship between log E and M, log E = 9.4 + 2.14 M− 0.54 M^2, which is valid for local magnitude [13]. In the book [4] (p. 366) a similar formula is given with somewhat different coefficients— log E = 9.9 + 1.9 M_L−0.024 M_L^2—which agrees better with equation (1). But this does not change the principal point. It is important that Carder and Cloud, in using the squared relationship, make a valid distinction between the local magnitude M_L and the unified magnitude $\underline{m}$, for which the relationship between log E and $\underline{m}$, as is well known [2, 4], is different; it is linear: log E = 5.8 + 2.4 m. In subsequent work, assuming extensive absorption of the seismic waves near the explosion, these authors state that, beginning at a distance of 2 km, it is more suitable in making comparisons with earthquakes to use a value of M ≈ 4.0. But we will not here enter into a consideration of the basis for the indicated assumption of strong absorption.

Let us turn to the data on the magnitude of the Rainier shot, as described at the first conference of experts at Geneva in 1958. For this, the only underground nuclear explosion up to that time, data were presented only for local magnitudes in the close-in zone, and for only seven of the ten stations indicated in Table 1. Data for the Riverside, Woody, and Barret stations (noted by asterisks) were then missing. As explained by the American delegation later, at the second conference in December 1959, this was due to the "low precision resulting from small amplitudes" on the records of the Rainier shot at these stations. It is characteristic that all three of these added stations give smaller magnitudes (less than 4) than the other seven stations (4 and more). The values are especially small on the records of the Barrett and Woody stations.

Without considering the indicated three stations, the average magnitude value for the Rainier shot was determined to be $M_L = 4.27$. Approximately this same value, specifically $M_L = 4.25 \pm 0.4$, was that officially represented by the American delegation in 1958. It was used as a basis in computations at the first conference, the difference between M_L and $\underline{m}$ apparently not being taken into account.

In the first official document [10] on the "New Seismic Data," there were still no figures in the places marked by asterisks in Table 1. However, it was said that measurements of magnitudes for the Blanca and Logan shots made at all ten stations showed that "observations for the Rainier shot were made only at stations giving anomalously high magnitudes," and that, in connection with such comparisons, it was necessary to reconsider the magnitude of the Rainier shot and to decrease it to the value $M_L = 4.1 \pm 0.4$.

A similar value was given in the paper of Romney [9] (4.0, 4.05, or 4.1) and in this author's principal report at the second conference of experts. As we may see, this value agrees with the later and still somewhat lower average value obtained from all the American data for this shot as given in Table 1, with due consideration given to the supplementary values from the three stations having low values:

$$M_L = 4.06 \pm 0.13. \tag{2}$$

Here $\pm$ 0.13 is the standard (average squared) deviation in the average value of M_L; in the two preceding evaluations $\pm$ 0.4 is the standard deviation for a single determination.

We have pointed out below that, instead of "anomalously high" values of magnitude at the seven stations supplying data for the Rainier shot, we should speak of "anomalously low" values obtained at the three other stations, especially for the Barrett and Woody stations. We shall now temporarily assume that the value in (2) for the magnitude of the Rainier shot is valid, and shall use it in subsequent computations.

Let us turn to a discussion of the magnitudes of the Blanca and Logan shots. Observations on these explosions were made not only in the first, local, zone (as for the Rainier shot) but also in the second zone, the zone of diffraction, which extends approximately from 1100 to 2500 km, and also in the zone beyond, the third or distant zone, extending up to 10,000 kilometers from the epicenter. There is thus an abundance of material for analyzing and comparing the seismic effects of explosions and of earthquakes.

We shall first attempt to follow the reasoning of the American delegation on this problem.

In the "Work Report" [10] and in a later paper [9], and also in the principal report of Romney at the second conference, the individual average local magnitudes for the first zone were first computed as an approach to determining the average magnitude values for the Blanca and Logan shots; the following values were thus obtained (which are in agreement with the latter summary of December 14, 1959, see Tables 1 and 2):

average local magnitudes M_L $\begin{cases} \text{Blanca} & 4.76 \pm 0.13 \\ \text{Logan} & 4.44 \pm 0.12 \end{cases}$

and the values computed with data from the second and third zones were

average teleseismic magnitudes $\underline{m}$ $\begin{cases} \text{Blanca} & 4.84 \pm 0.11 \\ \text{Logan} & 4.44 \pm 0.21 \end{cases}$

Here the figures after the "±" sign represents the standard deviation for the mean values. The further course of the treatment is as given below (see [9]).

From the two average figures for the first zone and for the second and third zones together for each explosion, it may be seen that the average magnitudes for these explosions, determined by the two methods according to the two different scales, M_L and $\underline{m}$, are practically identical. From Gutenberg's point of view this coincidence should be regarded as unexpected, at least for earthquakes; his investigations show that determinations of the magnitudes of earthquakes on the M_L and $\underline{m}$ scales should differ at a magnitude of 4.5, as shown by Romney, approximately by one magnitude unit. "However, this discrepancy may not be significant because of the uncertainty of the relationship between the two magnitude scales" [9, p. 1497].

After this, a decisive step is taken: it is assumed that the local (Table 1) and teleseismic (Table 2) magnitudes represent a homogeneous (uniform) sequence of values. It is further tacitly assumed that deviations are absent in this sequence (all deviations are considered random); all observations are given identical weight, and an over-all average magnitude value for all three zones is computed, expressed in an undefined scale. The averages of all the zones of the so-called "nonscale" magnitudes are shown by figures [9, 10] that are in accordance with the values in Table 1 (for M_L) and Table 2:

$$\text{nonscale magnitude} \begin{cases} \text{Blanca} & 4.8 \pm 0.1 \\ \text{Logan} & 4.4 \pm 0.1 \end{cases}$$

where ± 0.1 is the standard deviation of the mean.

It is interesting to note that the above-cited disregard of the necessary correction, from the point of view of Gutenberg's investigations, consisting of one whole scale unit in converting M_L to $\underline{m}$, as indicated in [9], exists side by side in the official position of the American experts with the fact that they ascribed great significance to the change in magnitude of the Rainier shot from 4.25 to 4.1 (see [10]), i.e., a change of only 0.15. It is possible that an important factor here was the difference in sign of the two corrections and the consequences evolving from this: a negative correction of 0.15 corresponds to an increase in the number of earthquakes exceeding in magnitude explosions of the given yield by approximately fifty percent; i.e., it allegedly attests to the great difficulty of distinguishing explosions from earthquakes, whereas a positive correction of 1.0 would lead to a decrease in the number of such earthquakes, to approximately one-tenth the number, giving one grounds to conclude that it was considerably easier to distinguish explosions.

The failure of the American experts, in analyzing and discovering the causes of the "unexpected" (to them) agreement (see [9]) between the local and teleseismic magnitudes in the indicated experiment, can only be explained in this case, by the hypnotic effect of the very fact that numerical values obtained by different means were the same. As will be seen below, this coincidence is actually remarkable: if the American temporary seismic stations that recorded the Blanca and Logan shots had been distributed differently between the second and third zones, or even within the second zone alone (where the strong systematic variation in amplitude was observed), the agreements of values might not have occurred, and then the analysis outlined below would perhaps have been made by the American investigators themselves.

2. Analysis of the Magnitudes of the Explosions

Systematic Variation in the Magnitudes of the Explosions. Let us turn our attention to the relationship between the epicentral distance $\underline{x}$ and the first measured magnitude values M_L and $\underline{m}$ of the Blanca and Logan shots. For this purpose we shall use the mean values of M_L and $\underline{m}$ computed in Tables 1 and 2 from the three different zones; we shall also use Figs. 2 and 3, which give the proper graphs to show the relationship between magnitudes (M_L in the first zone and $\underline{m}$ in the second and third zones) and distance. In Figs. 2 and 3 the original data on magnitudes were presented tentatively, since all the primary measurements on magnitudes were fully comparable, i.e., in the sense of the American investigators.

It is pertinent to recall here that the magnitude of an earthquake is a value characterizing the seismic effect, strictly speaking, at the very focus of the earthquake, and therefore, in its proper sense, it does not depend on the epicentral distance $\underline{x}$. It is precisely from a consideration of this fact that seismologists began to construct magnitude scales. Calculations for each of these scales individually or for any corresponding combination of them should lead to equal magnitudes measured at any desired epicentral distances and expressed in units of a single system; this has to do, of course, only with a systematic variation of values, since random deviations in measured values from the means (random variations in magnitude) are possible, about which we will speak below. Under such circumstances the average magnitudes of earthquakes, particularly from measurements in the first, second, and third zones, should be the same within the limits of random variation. If we make use of the known differences between earthquakes and explosions as sources of seismic waves, then for explosions we should expect the magnitude to be independent of distance and should expect the mean values to be the same in the different zones, if, of course, they are everywhere expressed in the same system of units. This should be expressed graphically in Figs. 2 and 3 by only random deviation of the observed points from a relatively horizontal straight line intersecting the vertical axis at a point equivalent to the mean value.

Meanwhile, from Tables 1 and 2 it may be seen that, for both explosions—Blanca and Logan— the mean values of the nonscale magnitudes (in the sense of the papers [9, 10]) show marked differences in the indicated three zones. In the shadow zone (2) the mean value is approximately 0.5 less than in the distant zone (3). And in the local zone (1) this value is approximately 0.3 less than in the distant zone.

These differences are very graphically revealed on the curves in Figs. 2 and 3. Here one may also observe some important details in the distribution of corresponding points. The first thing that strikes the eye is the systematic pattern of deviations in the points of the second zone, the zone of diffraction: on the average the magnitudes decrease with distance into this zone from the boundary with the third zone, i.e., from right to left.

Although the differences in measured values of the magnitudes $\underline{m}$ in the second and third zones, seen in Figs. 2 and 3, are convincing to the eye, we may make the simplest quantitative statistical analysis of these differences, assuming conditionally that the deviations of individual values from the mean values in both zones are random, but that the mean values are identical or that they differ among themselves. The corresponding computations were made by V. F. Pisarenko.

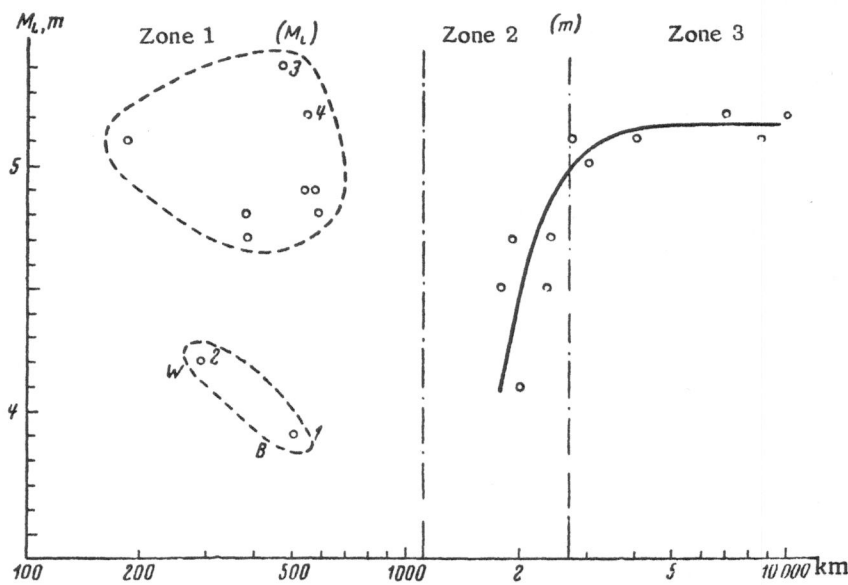

Fig. 2. Relationship between measured magnitudes M_L and $\underline{m}$ and epicentral distance $\underline{x}$ for the Blanca shot. 1) Barrett, 2) Woody, 3) Mt. Hamilton, 4) Palo Alto.

Let us consider, as is generally done in statistics, that the actual measured magnitudes in the second and third zones represent a selection of limited range from some general population corresponding to an infinitely great number of observations. Let us designate the selected mean of the magnitude values from the second zone by $\bar{x}_2$, and from the third zone by $\bar{x}_3$. And let us designate the corresponding mean values of the general population by m_2 and m_3. Let us assume that the magnitude values in the second and third zones are distributed normally and are characterized by the parameters (m_2, σ_2) and (m_3, σ_3). In this consideration σ_2^2 and σ_3^2 are the known dispersions of the general populations for the second and third zones. For our purposes it may be stated approximately that

$$\sigma_2 = S_2,$$
$$\sigma_3 = S_3; \tag{3}$$

in which S_2^2 and S_3^2 are the selective dispersions [as another variant we consider below a different hypothesis in place of (3)]. In the proposed relationship (3), the value $\bar{x}_2 - \bar{x}_3$ will have normal distribution with the parameters*

$$\left(m_2 - m_3, \ \sqrt{\frac{S_2^2}{n_2} + \frac{S_3^2}{n_3}} \right),$$

where n_2 and n_3 are the numbers of observations in the second and third zones.

Let us now consider, if it may be permitted, what values of magnitude observed in the second and third zones for the Logan and Blanca shots are in agreement with the hypothesis $m_2 = m_3$, i.e., what measured values of magnitude for these explosions, in both zones, may be considered identical within the limits of random variation.

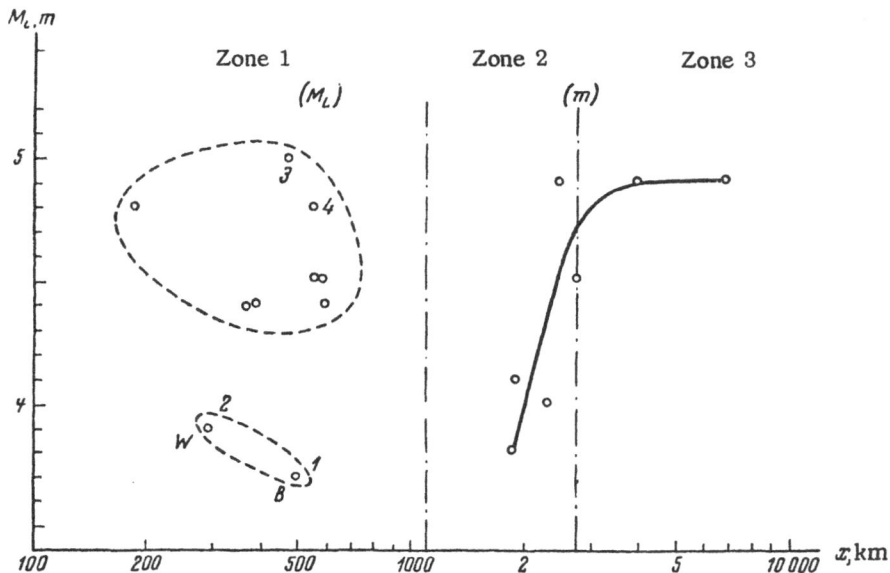

Fig. 3. Relationship between measured magnitudes M_L and m and the epicentral distance x for the Logan shot. 1) Barrett, 2) Woody, 3) Mt. Hamilton, 4) Palo Alto.

For the Logan shot we obtain the following

$$\bar{x}_2 = 4.16; \quad \bar{x}_3 = 4.78; \quad \sqrt{\frac{S_2^2}{n_2} + \frac{S_3^2}{n_3}} = 0.24;$$

$$\frac{\bar{x}_2 - \bar{x}_3}{\sqrt{\frac{S_2^2}{n_2} + \frac{S_3^2}{n_3}}} = -2.58.$$

Meanwhile, the confidence interval, embracing 98% probability, is (−2.33, +2.33). We see that the measured value −2.58 is not within this interval. Thus, the hypothesis of $m_2 = m_3$ has a significance level of no more than 2%; i.e., the probability is small and the hypothesis should be discarded.

For the Blanca shot we find the following equations

$$\bar{x}_2 = 4.5; \quad \bar{x}_3 = 5.1; \quad \sqrt{\frac{S_2^2}{n_2} + \frac{S_3^2}{n_3}} = 0.105;$$

---

*See N. V. Smirnov, I. V. Dunin-Barkovskii. Brief Course in Mathematical Statistics for Technical Supplements [in Russian], Gos. izd. fiz.-mat. lit., Moscow, 1959.

$$\frac{\overline{x}_2 - \overline{x}_3}{\sqrt{\dfrac{S_2^2}{n_2} + \dfrac{S_3^2}{n_3}}} = -5.71.$$

The confidence interval, embracing 99.9% probability, is (−3.3, +3.3) (see the same reference). The observed value of −5.71 is not within this interval. Thus, in this example, the hypothesis $m_2 = m_3$ has a significance level of 0.1%, even smaller than in the preceding example, and it should be unconditionally rejected. The more definite results for Blanca may be explained to some degree by the greater number of observations (11) than for Logan (7).

In place of the proposed relationships in (3), another variant for computation might have been proposed; namely, that

$$\sigma_2 = \sigma_3. \tag{3'}$$

Then the properly normed value

$$\frac{\overline{x}_2 - \overline{x}_3}{\sqrt{n_2 S_2^2 + n_3 S_3^2}}$$

has Student distribution (see the same reference). As computations have shown, in this example the significance level, such that the hypothesis $m_2 = m_3$ is rejected, is 2% for Logan and 0.1% for Blanca; i.e., these values are the same as with the first variant for computation with the proposed relationship in (3).

Thus, a quantitative statistical evaluation leads one to conclude that the differences observed in Figs. 2 and 3 for the measured magnitude values of the Logan and Blanca underground nuclear explosions in the second and third zones cannot be explained by random scattering.

The systematic decrease in magnitude of the Blanca and Logan shots in the second zone relative to the third zone is not related to the choice of magnitude scale: the same scale m was used in both zones.

There remain the assumptions that the observed phenomenon is due either

a) to "random" local peculiarities of the seismic stations occurring in this zone, or

b) to regular systematic differences between waves from earthquakes and from explosions related to geophysical causes of formation in this zone, as a special zone of diffraction, and to differences between earthquakes and explosions, as inhomogeneous sources of seismic waves.

There is but scant material, unfortunately, to judge the value of point (a). The question might be shown in better light had observations been made at those stations on earthquakes as well as on explosions. However, this was not done, and for this reason there is a serious deficiency in the technique of the seismic investigations carried on during the Hardtack series of tests. Nevertheless, it is sufficiently clear that the observed variation of magnitudes is not random and does not have narrow local causes. On the average the variation is the same for both explosions, despite the fact that the sites of most of the stations were changed between explosions. It should be noted that this total pattern is manifested in a rather extensive region, extending along a line for approximately 1000 km.

More definite judgments may be made for point (b). On the one hand, it has long been known from seismology that there exists a shadow zone in which seismic signals are complicated and diffused. In this zone clear first arrivals of P waves having especially high frequency disappear almost entirely. Gutenberg has stated, and this has been confirmed by later investigations, that this is due to the presence, under the layer with high seismic velocity directly beneath the Mohorovicic discontinuity, of a layer with somewhat lower velocity; but deeper yet the velocity again begins to increase. This structure must produce the diffraction phenomena that are observed at some distance from the source. But we shall not go into details of geophysical explanations. For us it is important here to note merely that a zone of diffracted signals exists and that in this zone the signals are rather poor in high-frequency components.

On the other hand, it is also well known [19, 20] that an explosion gives off relatively higher-frequency waves than an earthquake having the same seismic energy at the source. This results from the very restricted local character and the short-period action at the focus of an explosion as compared with the focus of a comparable earthquake. By focus we mean the zone of intense disruptive and nonlinear movement, outside of which stresses and deformations in the seismic waves are continuous and, basically, linear.

From a comparison of both points it immediately follows that in the zone of diffraction, which is especially unfavorable for the passage of high-frequency waves, the seismic effect of an explosion, rich in high-frequency components, should be weaker than the effect of a comparable earthquake. This naturally has a direct bearing on the magnitudes.

Thus, the peculiar behavior of magnitudes observed for the explosions in the second zone has a very natural geophysical explanation. More than this, it would be extremely strange if the indicated differences were not observed in the zone, differences associated with the different frequency spectra for explosions and for earthquakes.

And it is precisely with such an unaccountable singularity that we are faced when we go on to an examination of the magnitudes in the first (local) zone, which are regarded by American investigators, from the viewpoint that the local M_L and teleseismic $\underline{m}$ magnitudes may be thought to constitute an alleged "homogeneous" population $M_L \equiv m$.

Actually, since explosions produce higher frequencies than earthquakes, waves from explosions should die out with distance more rapidly than those from earthquakes. In the distant (third) zone, where all observed waves are of

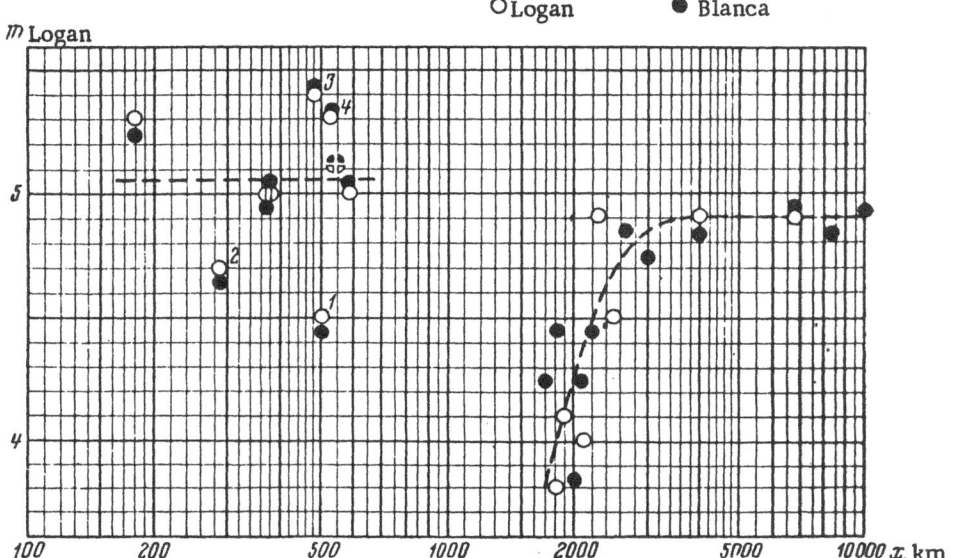

Fig. 4. Generalized systematic relationship between the unified magnitude $\underline{m}$ and the epicentral distance for Logan and Blanca. The figures designate the stations indicated in Figs. 2 and 3.

relative low frequency because of previous absorption of the high-frequency components, the signal from an explosion should be weaker than the signal from an earthquake. For this reason, if both signals are equal in intensity in the third zone, the level of the signal from the explosion ought to be higher in the first zone than the signal from the earthquake. The same argument applies to magnitudes. However, as we have seen (Tables 1 and 2 and Figs. 2 and 3), the relationship is reversed for the "nonscale" magnitudes of the Blanca and Logan shots.

This apparent contradiction disappears when one ceases to ignore the difference between local and teleseismic magnitudes, as indicated by Gutenberg, and reduces all magnitudes to units of the unified scale $\underline{m}$ (see the proper column in Table 1). As we see, the magnitudes $\underline{m}$ for explosions actually prove to be, as should be expected from spectral considerations, somewhat greater in the first zone than the magnitudes $\underline{m}$ in the third zone. The difference is not large, amounting at best to but 0.2 of an $\underline{m}$ unit; this is approximately twice the standard deviation for determinations of the mean magnitudes separately for the first and third zones. The cited analysis attests to the approximate correctness of Gutenberg's correlation of the M_L scale to the $\underline{m}$ scale [3, 4] and contradicts the "uncertainty" mentioned by Romney [9, p. 1497] as an excuse for his procedure in averaging all the values.

Figure 4 shows the generalized relationship between the unified magnitudes $\underline{m}$ and the epicentral distance. Here the data for Logan and Blanca are tentatively combined and reduced to the level of Logan. This was done in the following manner: the magnitudes of the Logan shot were left unchanged, and 0.26 was subtracted from each magnitude of the Blanca shot, this value being the difference between the mean values of $\underline{m}$ for both shots in the local zone where observations were made at the same ten stations. As may be seen from Fig. 4, in this case for the points $\underline{x}$, $\underline{m}$ values form automatically a single sequence for both shots in the zone of diffraction and in the distant zone ($\underline{x}$ = 1700-10,000 km). It is of interest to note that an attempt at such a combination, which we made earlier for the magnitudes M_L and $\underline{m}$, not reduced to a single scale, was not successful. It was demonstrated that when the points $\underline{x}$ are combined (M_L for the local zone at point $\underline{x}$ and $\underline{m}$ for both blasts from the two other zones) a systematic divergence occurs, especially noticeable in the distant zone: the points for the Logan shot occur above those for the Blanca shot. This fact may be considered as yet another supplementary witness in favor of Gutenberg's correlation of the M_L and $\underline{m}$ scales and in favor of its close applicability to explosions.

In passing, it may be noted that the dashed averaging (trend) line on Fig. 4 may be used to determine the corrections for measured magnitudes $\underline{m}$ of explosions in order to obtain a single magnitude value reduced to the level of any specific zone, such as the local zone.

It should be noted that a computation of the spectral difference between explosions and earthquakes does not yet fully cover the essential aspect of the question concerning a comparison of the two sources, in particular in regard to their magnitudes in the local zone. There is also a difference between explosions and earthquakes in the distribution of energy among the different types of waves at the focus: between longitudinal and transverse waves and between body and surface waves. There is also a difference in the distribution of energy among the different components of the waves.

It is thus natural to expect that for explosions the ratio of energy of longitudinal waves to energy of transverse waves is greater than for earthquakes. Furthermore, for any particular total energy at the focus, the energy of high-frequency surface waves should be greater for explosions than for earthquakes (because of the high-frequency character of an explosion and because of the shallow depth of the focus), and the energy of the low-frequency surface waves should be less (because of the weakness of the low-frequency part of the spectrum in explosions). In this connection, the computation of relative magnitudes of explosions and earthquakes in the local zone for "shear waves," including both transverse and surface waves, is very formal, and it is difficult to compare such values purely by empirical means with magnitudes computed from longitudinal waves.

The wave pattern during explosions should generally be simpler than that for earthquakes, because of the short duration of the explosion itself and also because of the shallower depth of the focus, where it is difficult for any waves reflected from the earth's surface, which would not be superimposed on the signal of direct waves, to form in the epicentral zone.

Explosions and earthquakes should differ in the relative intensities of the transverse and radial components of the waves, especially in transverse waves (both body and surface): for explosions this ratio should be smaller because of approximate axial symmetry of the source. This should be especially perceptible at rather low frequencies, when the influence of local horizontal inhomogeneous media is obliterated. Thus the intensity ratio of the low-frequency Love and Rayleigh waves for explosions should be, on the average, much less for earthquakes.

There is already some observational material on all these differences, but it should be investigated with special care, in order to work out composite criteria for differentiating explosions from earthquakes, more effective criteria than the single one recognized at the present time, i.e., the direction of the first movement.

A comparison of seismic effects, particularly the magnitudes of explosions and of earthquakes, should be made for all waves and all components for which it is assumed there are differences. If explosions are to be differentiated from earthquakes only by the direction of the first movement in the longitudinal waves, then the magnitudes should be computed for these waves not only in the local zone but in the others as well. For this purpose a magnitude scale is needed in the local zone based on the first longitudinal waves.

Thus, we see that the magnitude of an explosion, computed by methods worked out for earthquakes and used for comparing the seismic effects of explosions and earthquakes, proves to have a value that depends on the indices for which the comparison is made. If we confine our comparison of magnitudes of "shear" waves to the local zone and of amplitudes and periods of longitudinal waves to the diffraction and distant zones (with the computation of

M_L and $\underline{m}$, respectively), as was done in the treatment of the seismic data for the Rainier, Logan, and Blanca shots, and if, in all this, we express all measured magnitudes in a single system of $\underline{m}$ units in the unified scale of Gutenberg, we then obtain results that indicate that the magnitude of an explosion depends primarily on the epicentral distance. This has the greatest significance in the first (local) zone. In the distant (third) zone the magnitude is apparently somewhat less. In the diffraction zone (second), it is markedly less and changes in relation to the site within the zone.

Variation in the Magnitudes of Explosions. Until now we have considered the "systematic" side of the question concerning magnitudes of explosions. Let us now turn our attention to the "random" side: the variation in different computed values of magnitude relative to some average values. From Fig. 4 we see that this may be done without reservation only for the first zone ($x \leq 1000$ km) individually and for the third zone ($x \geq 2500$ km) individually, where the lines corresponding to average magnitudes are horizontal. In the second zone, where a well-defined systematic variation is observed, the variation should be computed not from a horizontal straight line but from an averaged rather steep curved (trend) line.

Figure 5 shows the variation, for Logan and Blanca, of points of $\underline{x}$ and $\underline{m}$ from the averaged (trend) lines indicated in Fig. 4. One's eye is struck by the strong variation, on the one hand, in the first and second zones taken jointly, and the much smaller variation in the third zone, on the other. This relationship may be found also in the values of the standard deviations σ_m and σ_{M_L} for single determinations of the magnitudes $\underline{m}$ and M_L in the first and third zones (Tables 1 and 2). The deviations shown in Table 2 for the second zone obviously do not have the pattern of random variation, since they include a systematic component.

In the first zone the variation in the values of the magnitudes of M_L and $\underline{m}$, respectively are approximately equal to ± 0.4:

Blanca	± 0.45	± 0.31
Logan	± 0.40	± 0.28
Rainier	± 0.41	± 0.30

The value ± 0.40 was also given in the "Work Report" [10] for the entire set of magnitude determinations for all three zones, with no consideration of the differences in the magnitudes M_L and $\underline{m}$. This leads one to believe, of course, that we have to do with chance coincidence.

The standard deviation of ± 0.4, obtained from American observations on the magnitudes of the underground explosions, proved to be unexpectedly large, noticeably larger than the deviation generally obtained at good stations from observations on earthquakes. It seemed natural that the reverse relationship should obtain in the deviation values for explosions and for earthquakes: an explosion is a more symmetrical source than an earthquake and, for this reason, one might expect a more uniform spatial distribution of the seismic effect of an explosion. Attention was drawn to this strange circumstance as early as the beginning of 1959, at the time of the discussions on the question of "New Seismic Data" in the subcommittee for disarmament of the Committee on Foreign Affairs of the U. S. Senate under the chairmanship of Senator Hubert Humphrey. According to the testimony of L. Murphy, head of the Seismology Branch of the U. S. Coast and Geodetic Survey, similar deviations for earthquakes are generally ± 0.1 to ± 0.2 at good seismic stations in the USA and in other countries. The same order of deviation is cited in an example in Richter's book [4, Table 22-2 on p. 343]. It is interesting that the lowest value of M_L for some of the earthquakes in this example, as in our observations of the underground nuclear explosions (Table 1, Figs. 2-4) were at the Barrett and Woody stations (points 1 and 2 on Figs. 2-5).

The question arises: why is the variation in the computed values of magnitude from explosions recorded in zones nearer the focus greater than those recorded in the distant zone. At first glance this may appear paradoxical; it would seem that the best place to evaluate an event would be immediately next to it. However, this is not so in this situation. Actually, if the observations in the local and distant zones are uniformly reliable, the observational data from the first zone should be more subject to fluctuations due to structural details in the inhomogeneous upper layers of the earth through which the waves in the zone travel. It is well known that structural inhomogeneities of the earth decrease with depth. In addition, the fluctuating pattern of the waves increases near the epicenter because of higher-frequency components among the seismic waves in this zone, as compared with the distant zone, and, correspondingly, because of less smoothing through structural peculiarities of the medium through which the waves pass. It is widely known that smoothing is greater the lower the frequency (greater the period) of the waves.

By uniformly reliable observations we mean that the observations included in computations were not marred by microseismic noise, a point of considerable value, of course, in the distant zone, where the useful signal is weak. Apparently microseisms caused little interference in determining magnitudes in the given case.

The smaller variation in magnitude values for earthquakes, in comparison with data for explosions, may apparently result from the fact that the determination of magnitudes of rather strong earthquakes that are considered in world statistics is most commonly made from observations at stations situated at the most favorable sites for this distant (third) zone. In addition, it is possible that in the determinations of magnitudes of explosions at near stations, a suitable system of station corrections may have been worked out, associated in particular with the consideration of possible differences in frequency and, perhaps, of the possibility that the parameters of the apparatus were not computed with absolute precision.

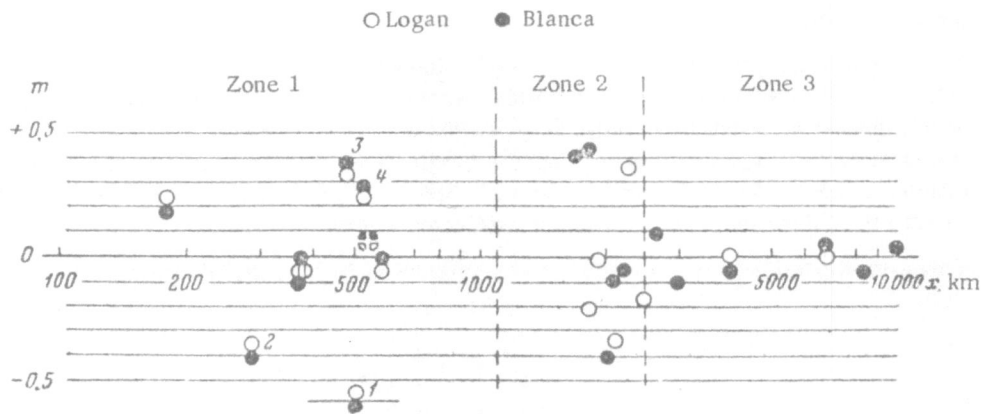

Fig. 5. Random variation of computed values of the magnitudes $\underline{m}$ for Logan and Blanca, measured from averaging (trend) lines (see Fig. 4). Figures designate stations indicated in Figs. 2 and 3.

Let us examine in detail some of the peculiarities of variation in values of the magnitudes $\underline{m}$ in the first zone. From a comparison of Figs. 2 and 3 and also of the black and white circles on Figs. 4 and 5, it may be seen that the deviations of individual values of magnitude from the mean for Logan and Blanca are almost the same for any particular station. The same is true in comparing data for all three shots (including Rainier). This comparison is shown in Table 3.

From a comparison of the standard deviations $\sigma_{\Delta m}$ of the differences Δm in magnitudes (Table 3) with the standard deviations σ_m of the values of $\underline{m}$ themselves (Table 1), it may be seen that the first are approximately one-third to one-sixth those of the second; i.e., the relative magnitudes have been determined much more precisely than the absolute values. This suggests that most of the observed deviations are associated with persistent conditions at the station (effect of local geology, installation of the apparatus, specific properties of the instrument, etc.) and not with random errors in determination, which may change from experiment to experiment (such as the effect of microseismic interference).

These deviations are especially large for the Barrett and Woody stations: the corresponding points 1 and 2 on Figs. 2-5 fall considerably below the denser grouping of points for the other stations. It is of interest to determine the degree of this discrepancy quantitatively. A treatment by least squares of the initial data— the American magnitudes M_L — supplies the following figures. The average local magnitude of Blanca for ten stations in the first zone, including Barrett and Woody, is $M_{L_{10}} = 4.79 \pm 0.12$, and for eight stations, excluding Barrett and Woody, is $M_{L_8} = 0.09$. Here the sign "±" gives the standard deviation from the mean values. We see that the exclusion of these two discrepant stations leads to a decrease in the error of average magnitudes from 0.12 to 0.09. If these points had not disagreed so markedly, the decrease in number of stations from 10 to 8 would have led, in general, to an increase in deviation of the mean values.

Let us now compute the deviations of magnitudes for Blanca at Barrett and Woody from the average value $M_{L_8} = 4.98$ for the other stations and compare these deviations with the standard deviation $\sigma_{M_{L_8}} = \pm 0.24$ for single

TABLE 3. Variations Δm in the Magnitudes of $\underline{m}$ for the Blanca-Logan and Logan-Rainier Shots

No.	Station	Δm	
		Blanca-Logan	Logan-Rainier
1	Tinemaha	0.2	0.4
2	Woody	0.2	0.3
3	Riverside	0.2	0.4
4	Pasadena	0.3	0.3
5	Mt. Hamilton	0.3	0.2
6	Barrett	0.2	0.3
7	Palo Alto	0.3	0.3
8	Berkeley	0.3	0.3
9	San Francisco	0.3	0.2
10	Mineral	0.3	0.0
	Mean variation	0.26	0.27
	Standard deviation:		
	single determination	± 0.05	± 0.12
	mean value	± 0.02	± 0.04

measurements at stations in this group after trebling this standard deviation. We obtain the following deviations for the stations:

$$\text{Barrett} \qquad 3.9 - M_{L_8} = -1.08$$

$$\text{Woody} \qquad 4.2 - M_{L_8} = -0.78$$

Trebling the standard deviation gives $3\,\sigma_{M_{L_8}} = \pm 0.72$; i.e., the deviations of both these points exceeds the trebled value of the standard deviation for the group of eight remaining points.

Let us employ, however, still another, milder, and, possibly, more valid evaluation. In a similar computation let us assume that the Barrett and Woody stations were included in the set of stations of the first zone, for which the average magnitude is determined. This leads to an approximation of the mean value of $M_{L_{10}}$ to values of M_L at these stations, and it increases the standard deviation for the entire group of points. As a result, we here obtain $M_{L_{10}} = 4.79$ and $\sigma_{M_{L_{10}}} = \pm 0.38$ and we then find the following deviations for the stations:

$$\text{Barrett} \qquad 3.9 - M_{L_{10}} = -0.89$$

$$\text{Woody} \qquad 4.2 - M_{L_{10}} = -0.59$$

Three times the standard deviation gives $3\,\sigma_{M_{L_{10}}} = \pm 1.14$.

In this last computation the deviation for Barrett and Woody falls within the trebled value of the standard deviation; it is about double the value, $2\,\sigma_{M_{L_{10}}} = \pm 0.76$, which, however, is sufficient to constrain us to use a certain caution when using data from these stations.

Professor Frank Press, director of the seismological station at Pasadena and supervisor of the Barrett and Woody stations, has reported to us that these stations lie on bed rock and are distinguished by a low noise level, and that, in general, they deserve every confidence. Keeping in mind his authoritative judgment, but admitting that there may exist at these stations some difficult-to-control undesirable deviations from the norm, let us apply for advice to Professor Hans Bethe for determining the mean value M_z in the first zone, using a procedure which includes the following: in computing $\overline{M_2}$ we exclude the values of the points (Barrett and Woody) with the smallest magnitudes, but at the same time we also exclude the two points with the greatest magnitudes (Mt. Hamilton and Palo Alto; points 3 and 4 on Figs. 2-5). Professor Bethe has accompanied this advice with the following very general and very instructive discussion. It may be assumed that all the observed deviations of M_L in the given zone are associated chiefly with peculiarities of geologic structure in regions of the individual stations and along the path the seismic waves traveled from the test site in Nevada. Negative and positive deviations are equally probably at all stations, in seismic energy and, correspondingly, in magnitude, from some average pattern. However, strong dissipation of energy is generally much more common than strong concentration; this latter is achieved only with great difficulty and within definite limits. Therefore, there is much better basis for expecting considerable deviation on the side of deficient expression of energy (such as at Barrett and Woody) than of great deviation on the side of over-expression of energy (which, for this reason, is not observed in these records).

Following the advice of Professor Bethe, having excluded two points from the top and two points from the bottom, we obtain a mean local magnitude of Blanca for the first zone from the remaining six stations of the zone

$$M_{L_\bullet} = 4.86 \pm 0.06.$$

This determination is somewhat higher than and twice as precise as the average of all ten stations, $M_{L_{10}} = 4.79 \pm 0.14$. However, it scarcely procures great formal precision. Admitting the possibility of a slight lowering of the determined value of $M_{L_{10}}$, we shall still adopt it as a basis for further computations.

The consideration given for the Blanca shot applies in equal degree to the other two, Logan and Rainier, for which, in general, the same distribution of M_L values is observed in the first zone. The conclusion that it is possible to use average magnitudes in the first zone, computed from all ten stations in this zone, applies, of course, both to local and to unified magnitudes.

Let us still take some note of the variation in magnitude values in the second zone, the zone of diffraction. This variation, as we have seen (Fig. 5), is on the average approximately the same as in the first zone. The epicentral distances in the second zone are greater but the frequencies here are lower, and, starting only from spectral considerations, it would seem that in this zone one might expect a great averaging of the propagation conditions of seismic energy and, correspondingly, a less random variation in magnitudes. However, there are other, stronger, factors altering this situation. The fact is that the intensity of the seismic waves in the second zone is capable of changing markedly with even slight changes in the composition or structure of the layers shielding and underlying wave channels (according to Gutenberg), the presence of which is connected with the very existence of this zone. The inconstancy of the seismic signal of explosions in the second zone cannot be surprising to seismologists, since this is a well-known phenomenon of earthquakes. Moreover, it is precisely this circumstance that makes it very difficult to develop a sufficiently precise or, more properly, sufficiently constant magnitude scale for earthquakes for the second zone. For this reason there has not been yet developed, in general, any magnitude scale for the second zone. It is known from seismology that the inconstancy increases the farther into the second zone one goes, a fact well shown by the variation of magnitude values for explosions (see Fig. 5): the clusters of points of x and m are scattered noticeably at increasing distances into the second zone from the boundary with the third zone.

Determination of Average Magnitudes. From all the preceding discussion it may be seen that the determination of average magnitudes for an underground nuclear explosion by averaging all the various measurements of local M_L and teleseismic m magnitudes, as cited in [9, 10], when these are not reduced to a single system of units, is nonsense. It is also difficult to determine average magnitudes on the basis of values expressed in a single magnitude scale for earthquakes, such as the unified scale m of Gutenberg or any other similar magnitude scale. The difficulties are due primarily to the systematic pattern of magnitudes of explosions relative to the epicentral distance.

For an explosion the magnitude worked out for earthquakes cannot, by its nature, serve as characteristic for the seismic effect of the focus alone; i.e., it cannot fulfill precisely the same function it generally fulfills for earthquakes. Still, the use of these magnitudes for explosions has a certain significance under definite circumstances. This significance involves a possible numerical comparison of certain aspects of the seismic effect of an explosion with the corresponding effect of an earthquake within definite epicentral distances, wave frequencies, and other conditions.

Despite these difficulties, it is still possible, in approximation, to speak of the magnitude of an explosion as having "on the average" a constant value within certain ranges of epicentral distances and for examined frequencies on the order of 1 cps. Judging from Fig. 4 this is fully admissible individually for the first and third zones, at least in the zones encompassed by the measured magnitudes. But if great precision is not required, it is possible to go even farther, neglecting the comparatively small systematic variation (about 0.2) in the magnitudes of the first and third zones, and to make a tentative determination of the average magnitude of an explosion for these two zones together. In doing this one should keep in mind that a change of 0.2 in m corresponds to a change in the TNT equivalent of an explosion two or three times that value.

In determining average magnitudes, the second zone, in view of the marked systematic variation of magnitude values relative to the epicentral distance, should naturally be excluded. This, however, by no means suggests that observations in this zone may not be used for determining average magnitudes in the other zones; it is important merely that, in making such computations, this circumstance of systematic variation in magnitudes should be kept in mind.

Average magnitude values $\underline{m}$ for Blanca, Logan, and Rainier computed individually for the first and third zones were shown earlier in Tables 1 and 2. We give now two variant computations of tentative "average" magnitudes for Blanca and Logan, made for the first and third zones jointly.

In the first variant we assume that all the individual determinations of magnitudes at stations in the first and third zones were the same. With this assumption for both explosions, we obtain, in correspondence with the numbers given in Tables 1 and 2

$$\text{Blanca} \ldots m_{B_1} = 5.23 \pm 0.06;$$
$$\text{Logan} \ldots \overline{m}_{L_1} = 4.97 \pm 0.08.$$

Here the sign "$\pm$" gives the standard variation $\sigma_{\overline{m}}$ of the mean. The corresponding standard deviations of single determinations of $\sigma_{\overline{m}}$ are ± 0.26 and ± 0.28.

In the second variant the over-all average for the first and third zones is computed as a weighted average of certain particular averages from the first and third zones, the weighted value $\overline{p}_i$ ($i = 1, 3$) in these determinations being inversely proportional to the square of the standard deviations of the particular means $\sigma_{\overline{m}_1}$ and $\sigma_{\overline{m}_3}$. In this computation we have for Blanca $\sigma_{\overline{m}_1} = \pm 0.10$ and $\sigma_{\overline{m}_3} = \pm 0.03$, and for Logan $\sigma_{\overline{m}_1} = \pm 0.09$ and $\sigma_{\overline{m}_3} = \pm 0.13$ (see Tables 1 and 2). As a result we obtain for the two explosions

$$\text{Blanca} \ldots \overline{m}_{B_2} = 5.14 \pm 0.04;$$
$$\text{Logan} \ldots \overline{m}_{L_2} = 4.95 \pm 0.10.$$

Here the standard deviations of the over-all means are computed by the formula

$$\sigma_{\overline{m}} = \pm \sqrt{\frac{\overline{\Sigma p_i \sigma_{\overline{m}_i}^2}}{\Sigma p_i}}.$$

In comparing the results of both variants, we see that they are very similar. If we take into account, furthermore, the very tentative character of the computation for the mean, in view of the systematic difference in the averaged values, we obtain for both shots, as our final result, the following mean values for the first and third zones taken together, the values being rounded off to 0.05:

$$\text{Blanca} \ldots m_{1,3} = 5.2 \pm 0.1;$$
$$\text{Logan} \ldots m_{1,3} = 4.95 \pm 0.1. \tag{4}$$

The difference between these values, 0.25, is almost the same as the more precise value 0.26 ± 0.02, which was previously determined from differences of magnitude at stations in the first zone that had records for all three shots (Table 3). By adopting $m_{1,3} = 4.95$ for Logan and computing the average Logan-Rainier difference to be 0.27 ± 0.04 (see Table 3), we obtain for Rainier the following magnitude value for the first and third zones (as in the preceding, rounded off to 0.05):

$$m_{1,3} = 4.7 \pm 0.1. \tag{5}$$

Let us note that for the first zone separately the average magnitudes for all three shots were approximately 0.1 greater, but for the third zone they were approximately 0.1 less than in equations (3), (4), and (5).

3. Relationship between Magnitude and Yield of an Explosion

In order to determine the number of earthquakes exceeding in magnitude an underground nuclear explosion of a given yield, it is necessary first to establish the general relation m (Y) between the yield of an explosion Y (in kilotons of TNT equivalent) and its magnitude $\underline{m}$.

<u>History of the Problem.</u> At the Geneva conference in 1958, when the experts presented actual data on but the single underground nuclear explosion (Rainier), there was no possibility of defining this general relationship from the data merely of the nuclear explosions themselves. For this, data from tests with detonations of chemical explosives were used, in addition to some theoretical considerations.

According to Carder [12, p. 1474], earlier observations on a number of chemical explosives, showed that the amplitude A of the ground because of seismic waves from an explosion varied as the $^1/_2$ to 1 power of the size of explosive charge Y. Since the magnitude <u>m</u>, for the same wave frequency, is proportional to log A (see, for example, [4]), the relationship between <u>m</u> and Y may be written in the following general form:

$$m = C + n \log Y,$$

(6)

where, according to Carder, at least for chemical explosives, n = 0.5-1.0.

In 1957, during preparation for the Plumbbob operation, two underground detonations of chemical explosives, with yields of 10 and 50 tons, were set off in the region of the later Rainier shot [12]. A comparison of the ground amplitudes A from these explosions led to the conclusion that $A \sim Y^{3/4}$, which corresponds to equation (6) with the coefficient n = 0.75. This value, as we see, lies within the above-indicated limits of 0.5 and 1.0.

At the conference of experts in Geneva in 1958, in establishing values for the parameters <u>n</u> and C in the general formula (6), a value of $^2/_3$ was adopted for <u>n</u> on the basis of some theoretical considerations and of experimental data for detonations of chemical explosives; this figure also lies within the limits of 0.5 and 1.0. The data of Rainier permitted only one parameter of C in equation (6) to be established.

The Hardtack II operation, with its large nuclear blasts of Blanca and Logan and the small Tamalpais and other explosions, in addition to the Rainier data, permitted refinement of the relationship in (6) from observations on the nuclear blasts themselves.

In the "Work Report" [10] it was pointed out, relative to the "New Seismic Data," that "the amplitude varies as the first power of the yield of the explosion within the ranges of 0.1-23 kilotons" according to the Hardtack tests; i.e., it is alleged that we obtain, in equation (6), n = 1.0. An explanation of the manner in which this value was obtained is given in a paper by Romney [9, Fig 3 on p. 1492]. It is shown that, to establish this relationship, use was made not of mutually agreeing determinations of explosion magnitudes but of relative amplitudes on records obtained at only some of the stations: at the temporary stations and at but one permanent station (Tinemaha, although as may be seen in Table 1, records were obtained from 10 permanent stations). The figure in Romney's paper [9] gives but one Tinemaha point for the Rainier shot. The averaging (trend) line, with a slope of n = 1.0, was drawn through a set of points for the large explosions (Blanca, Logan, and Rainier) and for two small explosions (Tamalpais and Neptune), although, as is known, Neptune was not essentially an underground explosion, since it produced extensive ejection of material into the atmosphere; for this reason, the seismic effect of this blast was minimized. In support of using the slope n = 1.0, Romney cites a theoretical discussion on this matter from the paper [14]. As a result, he finds for his magnitude M ("everything equal for whatever scale") the relationship

$$M = 3.65 + \log Y.$$

This same relationship, namely M = 3.65 ± 0.3 + log Y, was also adduced in the "Work Report" [10].

From the description of how this relationship was obtained, in [9], one can do nothing but doubt its validity. Coming to an analysis of the magnitude-charge relationship on the basis of more complete experimental data, let us first examine this question in regard to the large explosions: Rainier, Logan, and Blanca. For these blasts there is much more reliable observational material. Because of this, these explosions are of much greater interest in the problem of international control. We shall conclude this topic, however, with some discussion of explosions of smaller yield.

<u>The Magnitude-Yield Relationship for the Large Explosions.</u> Equation (6), with C and <u>n</u> being constant, does not at all represent a uniquely possible form of the general relationship between the yield of Y of an explosion and its magnitude <u>m</u>. More than this, there are indications that, in different ranges of yield Y, the slope n = Δm /Δlog Y on the graph showing the function <u>m</u> (Y) may change. A theoretical consideration leads one to conclude that when Y is small the slope should be near <u>n</u> = 1, but that when Y is large the slope should approach n = $^2/_3$. This theory does not yet give any indication of precisely what values might be expected as <u>n</u> changes from 1 to $^2/_3$. We should also keep in mind that the experimental data on chemical explosives show that the slope may be even less than $^2/_3$, dropping in some cases to as little as 0.5.

With the theory in such an inconclusive state, it is all the more important to establish first the proper empirical relationship, and it is precisely this goal that is given preference at the present time.

Since the chief uncertainty in the examined problem is the value of the parameter n, it is advisable for investigators to use values of magnitude differences (Table 3). It should be kept in mind that these differences were determined much more precisely ($\pm$ 0.02-0.04) than the absolute magnitudes ($\pm$0.1). In constructing a graph of m = f(log Y) within a limited range of yields—from 1.7 kilotons of Rainier to 19 kilotons of Blanca— it is clear that the curve of the averaging (trend) line may be replaced by a straight line.

The graph of the desired relationship is shown in Fig. 6. Here the segments passing through the points of Rainier and Blanca show the standard deviations in determination of the corresponding magnitude differences (see Table 3). In accord with Fig. 6 and with the above-cited precision evaluations, we obtain.

$$m = 4.6 \pm 0.1 + (0.50 \pm 0.06) \log Y. \tag{7}$$

This equation is also a solution to the problem concerning the relationship between magnitude m of an underground nuclear explosion and its yield Y (in kilotons) within the range of yields from Rainier to Blanca. It is perfectly clear (see Fig. 6) that the existing pattern may be extrapolated, with no great error, to the nearby environs of this range, specifically in the approximate range from 1 to 20 or 25 kilotons.

It may be of interest for comparative purposes to introduce the following equation for the relationship, similar to (7), obtained for local magnitudes of explosions (given with no accuracy evaluations):

$$M_L = 3.9 + 0.7 \log Y. \tag{7'}$$

Thus, we see that, on the basis of experimental data for the large nuclear explosions, the examined parameter of slope proves to have a value of n = 0.5 for the magnitudes m, which is much different from the value of 1.0 given in the "Work Report" [10] and in Romney's paper [9]. Moreover, even for local magnitudes M_L, which are approximated by the "nonscale" magnitudes of both papers [9, 10], the slope proves to be only 0.7, i.e., near $^2/_3$ or $^3/_4$, or near those values that were apparently adopted during the Geneva computations in 1958.

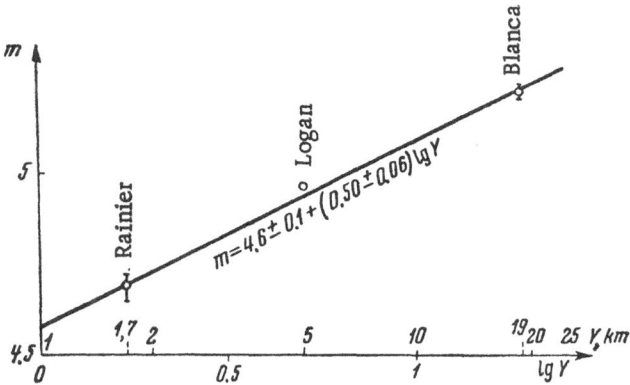

Fig. 6. Average relation (for first and third zones) between yield Y of an explosion (in kilotons) and its magnitude m, for the large explosions (Rainier, Logan, and Blanca).

It is possible that in the American treatment of "New Seismic Data" some role was played by the circumstance that an increase of n to 1.0 leads to a considerable increase in the computed annual number of earthquakes exceeding in magnitude nuclear explosions with yields approximately up to 20 kilotons, as compared with the previous Geneva evaluations (see [10]), and leads to development of the view that there is much greater difficulty in recognizing such nuclear explosions among the "tremendous" number of earthquakes.

Let us turn our attention to the fact that the comparatively small value of n = 0.5, which we obtained for nuclear explosions in the given range of yields, approximately from 1 to 20 kilotons, corresponds to the minimum value of n indicated by Carder for detonations of chemical explosives (n = 0.5-1). It is possible that, in this connection the values for chemical explosives used by Carder were on the average of lower seismic* yields than the nuclear blasts of Rainier, Logan, and Blanca. Actually, with an increase in yield of an explosion, the value of n apparently tends to decrease, and with a decrease in yield, n tends to increase. Thus, the results obtained by us for the large underground nuclear explosions are apparently in complete agreement with the existing experimental data for detonations of chemical explosives (on the average of lower seismic yield).

A Comparison with the Small Nuclear Explosions. In the paper by Romney [9] it was shown that the local magnitudes M_L for the small explosions of Tamalpais, with a yield of 0.072 kiloton, and of Neptune, with a yield of 0.090 kiloton, could be determined at but one station, Tinemaha, where the magnitudes of all three large explosions were also determined. The appropriate data are shown in the columns for M_L in the first line of Table 4. Data on the magnitude of the third small explosion, Evans, are not generally provided in the American reports, apparently because of its "mystery."

TABLE 4. Computation of the Magnitudes M_L and m for Tamalpais and Neptune

Data	Blanca		Logan		Rainier		Tamalpais		Neptune	
	M_L	m	M_L	m	M_L	m	M_L	m	M_L	m
Tinemaha	5.1	5.52	4.8	5.32	4.2	4.89	3.1	4.09	2.9	3.93
Romney [9]							2.6	3.71	2.4	3.55
Report of December 14, 1959							2.65	3.75	2.45	3.59
Average for first zone (see Tables 1 and 2)	4.79	5.30	4.44	5.04	4.06	4.77	2.82	3.88	2.62	3.72

Romney believes [9] that the single determination of magnitudes for the indicated two small explosions at the Tinemaha station should not be trusted, and he gives lower values of magnitude for Tamalpais and Neptune, specifically 2.6 and 2.4, respectively (see the second line in Table 4). According to [9] these values were obtained by comparing the amplitudes and periods for the indicated small explosions and for Logan. Similar magnitude values for these explosions were presented in the American list of December 14, 1959: 2.65 for Tamalpais and 2.45 for Neptune (see line 3 of Table 4). Although the reports of the American experts, by tradition, did not give an account of what magnitude scale is represented by these numbers, it is easy to surmise that the numbers must belong to the local scale M_L, since they were obtained solely by comparisons of observations in the first zone.

In going on to an analysis of these data, we shall first reduce all the indicated magnitudes of M_L to the unified scale according to equation (1) or by the corresponding nomogram in Fig. 1. These figures are shown in the columns for m in Table 4. Further, in considering that the magnitude differences Δm, as we saw for Rainier, Logan, and Blanca, are more accurate than the absolute values of m, we shall also determine differences in the present computation. We shall designate them by the first letter of the name of each explosion.

For all three existing measurements of the Tamalpais-Neptune magnitude difference, the results are identical and give $m_{TN} = 0.16$. A firm relationship between the magnitudes of these two explosions has also been established by the same data. Now, by using the first line of Table 4, we determine the magnitude differences between Tamalpais and the three large explosions. The values obtained are $\Delta m_{BT} = 1.43$, $\Delta m_{LT} = 1.23$, and $\Delta m_{RT} = 0.80$ (corresponding to differences for local magnitudes of 2.0, 1.7, and 1.1; our value of 1.7 for the Logan-Tamalpais magnitude difference is near the value of 1.8 given for this difference in the report of December 14, 1959). Finally, in considering the difference Δm obtained from the average magnitudes m in the first zone for Blanca, Logan, and Rainier (Table 1), we obtain magnitude values for Tamalpais of 3.87, 3.81, and 3.97 reduced to the values for these three explosions; the average thus obtained for Tamalpais is then $m_T = 3.88$. In considering the difference $\Delta m_{TN} = 0.16$ for Neptune, and rounding off to 0.1, we get the final values for the first zone:

$$\text{Tamalpais} \quad m = 3.9$$
$$\text{Neptune} \quad m = 3.7$$

---

* The seismic yield Y of a nuclear explosion, where the value of Y is given in kilotons of TNT equivalent, proves to be noticeably less than for chemical explosives of a charge of Y kilotons.

The magnitudes $\underline{m}$ for these explosions before rounding off and also the values of M_L computed from then are given in the last line of Table 4.

Figure 7 gives a graphic comparison of the magnitudes $\underline{m}$ as a function of yield Y of nuclear explosions— large (Rainier, Logan, and Blanca) and small (Tamalpais and Neptune)— from observations in the first zone. An averaging (trend) line has been drawn through the points for the large explosions, using the previously determined slope n = = $\Delta m / \Delta \log Y = 0.5$; and through the point for Tamalpais a straight line has been drawn with the theoretical slope n = 1.0 for "rather small" charges Y. The point for Neptune is on one side, since this explosion, as indicated previously, cannot be classed in a single sequence with the other true underground explosions.

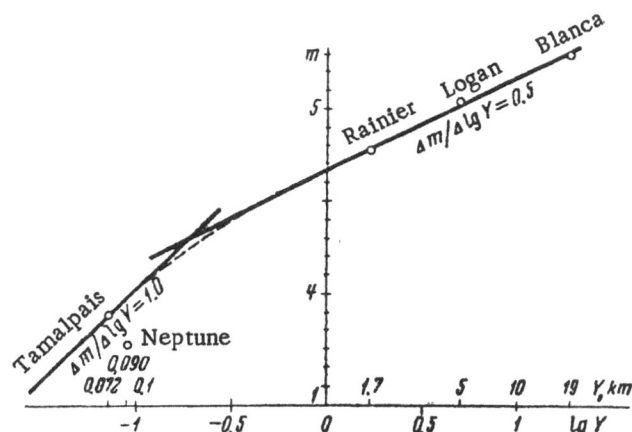

Fig. 7. Comparison of magnitudes $\underline{m}$ in the first zone for large (Rainier, Logan, and Blanca) and small (Tamalpais and Neptune) nuclear explosions. (Neptune is not considered an underground explosion.)

From Fig. 7 it may be seen that these two straight lines intersect in the region of yields Y < 1 kiloton (it may be mentioned in passing that this remains in force if the magnitudes M_L instead of $\underline{m}$ are plotted on the graph). It is natural to suppose that a change of slope on the graph of the relationship m (Y) occurs not in jumps but by very gradual change within the limits of some interval of Y values. This is illustrated in Fig. 7 in the dashed transitional curve. It may be seen that if this transition is extended into the zone of Y > 1 kiloton, the possibility of which is not excluded by the data, it cannot point basically to the course of the relationship m(Y) in the principal range of yields: from 1 to 20 kilotons.

It should be emphasized once again that the curve of Fig. 7 corresponds to the average magnitudes of $\underline{m}$ for the local (first) zone. Tentative "average" magnitudes of $\underline{m}$ for the first and third zones taken together (Fig. 6), as already mentioned, will be approximately 0.1 less. However, for such explosions as Tamalpais this has no significance, since it is hard to believe that systematic observations on such small explosions will be made at control points in the third zone.

4. The Number of Earthquakes Comparable to Explosions

At present the most reliable statistical data on the average annual number of earthquakes of different magnitudes for the world at large are given in the paper of Gutenberg [3]. These statistics are based on 50 years of instrumental records of earthquakes. The magnitudes are expressed in the unified system of units $\underline{m}$, the same units we have adopted in all the preceding computations for underground nuclear explosions.

The seismic effect of an underground explosion must be compared with the effect of shallow-focus earthquakes, the foci of which are in the earth's crust, at a depth (h) not exceeding 60 km. Earthquakes with deeper foci have peculiarities, known from the study of seismology, that permit one to recognize them and to allay any suspicion that the observed seismic phenomenon may be due to an artificial explosion. It is apparent that our present level of technology will not permit us to set off an explosion at depths greater than 10-15 km, to say nothing of depths of 60 km.

Analysis of Gutenberg's Statistics on Earthquakes. Gutenberg's data [3] on the number N of shallow-focus earthquakes ($h \leq 60$ km) are shown in the first and second column of Table 5. The third and fourth columns of the same table give the number N_Σ of earthquakes, computed from these data, for which the magnitudes m exceed the given value. Numbers (without query signs) of earthquakes are given to correspond with [3], the numbers being obtained by direct computation from records of a worldwide network of seismological stations; the numbers refer to rather strong earthquakes, with magnitudes $m \geq 4.2$. A question mark designates a number for weaker earthquakes, from data not completely embracing the modern records. These figures were obtained by recomputing for the entire world the records made in individual regions (California and elsewhere) having dense networks of stations.

Let us note first that the average magnitudes that we computed for the sequence of underground nuclear explosions—Rainier (m = 4.7), Logan (m = 4.95), and Blanca (m = 5.2)—are noticeably higher than the magnitudes in Gutenberg's list for which data are uncertain.

In Fig. 8 we show a graph of N_Σ (m), constructed from the numbers in the third and fourth columns of Table 5. A smooth curve is easily drawn through all the observational points, thus confirming, in general, the high quality of Gutenberg's statistics.

For a numerical control of the degree of "smoothness" of the computed relationship $N_\Sigma(m)$, and also for an explanation of some of the interesting properties of this relationship, a graphical relationship was computed and plotted (by means of numerical and graphical differentiation) for the relationship between density distribution of numbers of earthquakes and the magnitude $N' = - dN_\Sigma/dm$ (Fig. 9). The distribution of these points within the Gutenberg interval of reliable determinations $m > 4.25$ falls beautifully along a straight line with the slope

$$\beta = - \frac{d \log N'}{dm} = 0.99 \pm 0.01; \tag{8}$$

the only exception is a very small segment at the extreme right, for $m > 7$, which, as seen in Fig. 8, is based on a small number of observations of large earthquakes, at most about 10 per year. The value we obtained (8) is noticeably higher (above the limit of error) than the value indicated by Gutenberg [3], $\beta = 0.92$, which he obtained for $m < 7.1$ by direct examination of the number of earthquakes fitting in intervals of $1/10$ of a unit of the magnitude m.

TABLE 5. Average Annual Number of Shallow-Focus Earthquakes ($h \leq 60$ km) Throughout the World (N represents number of earthquakes in the indicated magnitude interval $\Delta m = m_i - m_k$; N_Σ represents number of earthquakes exceeding the indicated magnitude m)

$m_i - m_k$	N	m	N_Σ
$\geqslant 7.4$	3	7.35	3
7.0—7.3	11	6.95	14
6.2—6.9	80	6.15	94
5.5—6.1	400	5.45	494
4.9—5.4	1300	4.85	1794
4.3—4.8	4500	4.25	6294
3.5—4.2	30000?	3.45	36294?
2.0—3.4	800000?	1.95	836294?

On the left part of Fig. 9, i.e., where, according to Gutenberg, the statistical data are less reliable, the observational points deviate downward from our straight line, toward a smaller number of earthquakes. It is thought that this may possibly be due to imperfect calculation of the number of weak earthquakes. Actually, detailed observations on earthquakes made in a number of parts of the USSR in recent years attest to the linear character of the graph of $\log N' = F (\log E)$, where E, the seismic energy at the focus, a value linearly related to the magnitude m (see [13], and also [4], p. 365).

$$E = 5.8 + 2.4m, \tag{9}$$

persists through a wide range in energy E and, consequently, a wide range of magnitudes m as well [5, 6, 15-17].

The most detailed records, made in the Garm and Stalinabad districts of the Tadzhik SSR, showed that the slope of the graph of log N' = F (log E) is

$$\gamma = - d \log N / d \log E = 0.43 \pm 0.05. \qquad (10)$$

Using (9), from (10) we obtain for β the value

$$\beta = - d \log N'/dm = 1.03 \pm 0.12, \qquad (11)$$

which agrees, within the limits of error, with the value in (8), obtained from the reliable statistics of Gutenberg (β = = 0.99 corresponds to γ = 0.41). Let us note that our determination of (10) and (11) was obtained by investigation of earthquakes in the range of measured energy $E = 10^2$-10^{13} joules = 10^9-10^{20} ergs, i.e., in accord with (9) (where the value of E is given in ergs), within the limits of measured magnitudes of m = 1.3-5.9. This range overlaps considerably the span of unreliable data in Gutenberg's statistics.

The remarkable fact that this statistical pattern exists— the constancy of γ or β for various regions and over a wide range of measured seismic energies E or magnitudes m— permits us to extrapolate, with a considerable degree of assurance, data on the number of earthquakes in the world having small energies and magnitudes. It may be supposed that on this basis there is immediate possibility of determining, with rather high precision, the relative numbers of weak earthquakes in the world that may be directly recorded in the future and counted by improved operation of the entire expanded network of seismological stations.

The presence of this regularity makes it possible to obtain a solution as well for the following important problem: the establishment of spatial distribution, quantitatively expressed, of seismic activity in the earth.

At both conferences in Geneva, in 1958 and 1959, this problem was virtually ignored, since the discussions were confined to total numbers. Hence, it was only by crude determinations that the number of continental shallow-focus earthquakes was fixed at 50% of all shallow-focus earthquakes in the world. In the near future, however, it will undoubtedly become necessary to establish the proper number of earthquakes for individual countries and territories as well.

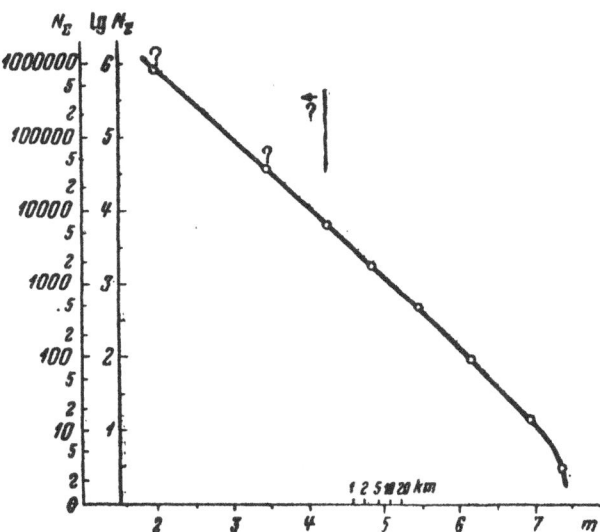

Fig. 8. Annual number N_Σ of shallow-focus earthquakes (h ≤ 60 km) on the earth exceeding in magnitude m the given value.

In the Soviet Union it has already been suggested [5] that the numerical seismic activity A may be determined as the annual number of earthquakes of any definite, fixed energy class, such as $E = 10^{10 \pm 0.5}$ joules, occurring in a restricted area, such as 10,000 km². In order to establish a value for A, records of earthquakes over a rather wide range of energies were used. From these data, a graph of frequency distribution of earthquake frequency of a particular energy was plotted for each rather large region; the result is a graph of "recurrence" log N' = F (log E). The value of A for an entire region and for the constituent segments is determined by the level of the averaging (trend) line on the recurrence graph along the N' axis, which is reckoned for the indicated fixed value of E. The spatial distribution of seismic activity A within an entire region is established on the assumption that γ = constant; it is shown on an isopleth maps representing equal values of A. Such a map permits one easily to compute the average number of earthquakes in any region, not only for the indicated fixed energy class, but also for all the remaining classes (excluding only the very rare catastrophic earthquakes), since the general relative law of frequency distribution for recurrence of earthquakes with particular energies is assumed to be known. Details of this procedure and some examples of applying the procedure have been published [5, 6, 15-17]; in these, all the statements concerning E may, of course, be applied approximately to m.

In the spring of 1958, at the session of the European Seismological Committee at Utrecht (Holland), the author reported on this question (published in the paper [15]), appealing to the seismologists of other countries to employ

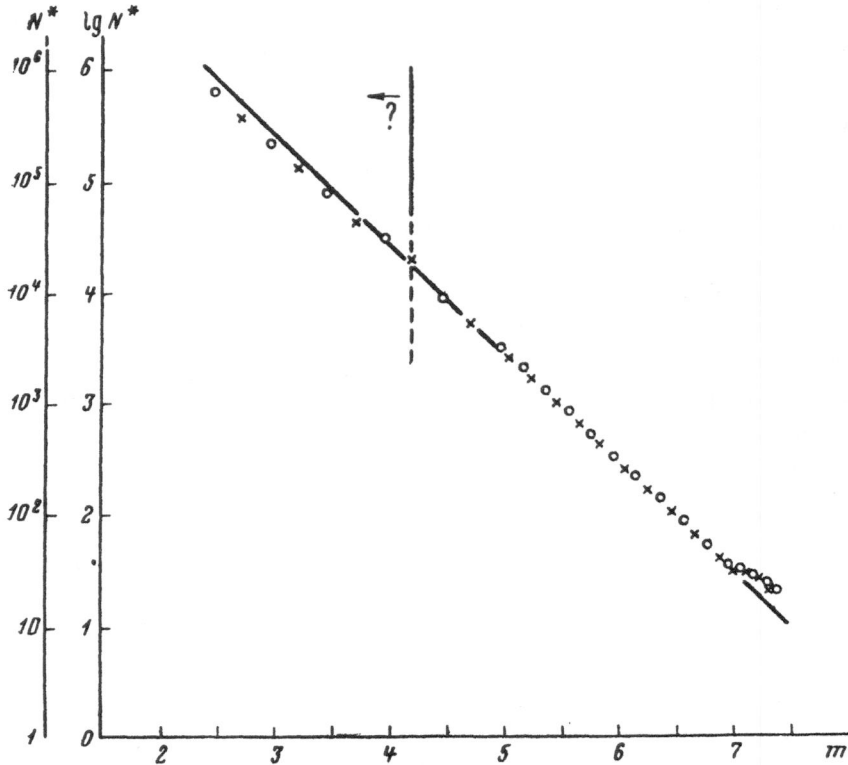

Fig. 9. Distribution by magnitudes $\underline{m}$ of the annual number N of shallow-focus earthquakes on the earth $N' = -\,dN_\Sigma/dm$. The circles represent graphical differentiation, crosses represent numerical differentiation.

the above-indicated method of accumulating suitable records, of treating the data, and of constructing maps of seismic activity for large regions, and then for the entire world. It seems to me that this proposal now requires special importance because of the problem of detecting underground nuclear explosion.

The Number of Earthquakes Exceeding in Magnitude Explosions of a Given Yield. In order to find the annual number of shallow-focus earthquakes N_Σ in the earth that exceed in magnitude $\underline{m}$ an underground nuclear explosion of a given yield Y, it is sufficient to compare the relationship between magnitude and yield m(Y) with the statistical number N_Σ (m) for earthquakes.

In Table 6 is given a comparison of the numbers of earthquakes N_Σ established by various means and at various times for some large values of Y of underground explosions: 1) from data of the Geneva conference of 1958; 2) from "New Seismic Data" in the Work Report" of January 5, 1959 [10]; 3) for numbers obtained by using the relationship $M_L(Y)$ for local magnitudes (7'), when it is erroneously contended, in the sense of the official position of the American delegation at Geneva in 1959, that the local magnitudes M_L coincide with the unified magnitudes $\underline{m}$, upon which Gutenberg's earthquake statistics are based [3]; 4) for the number N_Σ, obtained by comparing the average relationships m(Y), determined by us, for the first and third zones by equation (7) or by means of Fig. 6 and the curve in Fig. 8 for N_Σ (m), plotted in accord with Gutenberg's data [3].

The numbers with the "±" sign in the last column of Table 6 express the error in determining the number N_Σ of earthquakes because of the imprecision of determining average values of magnitude $\underline{m}$ within ± 0.1. The same numbers permit us to obtain some evaluation of the number of earthquakes that correspond to our observations individually in the first zone (with the "−" sign) and individually in the third zone (with the "+" sign).

In the "Work Report" [10] it is maintained that allegedly, according to new data, "the number of earthquakes per year equivalent to a given yield (of explosions) is approximately twice the number previously supposed," i.e., assumed at the conference in Geneva in 1958. In comparing the numbers in the second and third columns of Table 6 we find the same to be true, at least for explosions with yields between 1 and 5 kilotons. However, it may be seen

TABLE 6. The Annual Number N_Σ of Shallow-Focus Earthquakes in the Earth that Exceed in Magnitude Underground Nuclear Explosions of a Given Yield Y

Y, kiloton	No. of earthquakes, N_Σ			
	Geneva, 1958	"Work Report" of January 5, 1959	first zone on erroneous assumption that $M_L = m(?.)$	mean for first and third zone from our computations
1	10 000	26 000	13 000	3200 ± 600
5	3 800	5 800	4 000	1500 ± 300
10	2 400	3 000	3 000	1100 ± 200
20	1 500	1 600	2 000	800 ± 200

from the same table that the figures of N_Σ in the "Work Report" (third column) for explosions with Y = 1-5 kilotons are considerably, one and one-half to two times, greater than the figures (fourth column) deriving from the official position of the American delegation at the Geneva conference in 1959. The cause of this is due chiefly to an incorrect determination in the report [10] of the magnitude-yield relationship of an explosion, about which we spoke above (section 3).

The numbers in the fourth column, as might be expected, differ little from those in the report [10] for data at the Geneva conference in 1958 (second column). The difference that does appear is explained chiefly by the slight decrease in average magnitudes M_L due to inclusion of data from three new stations that showed lower magnitude values for the explosions, because of local peculiarities at these stations (chiefly the Barrett and Woody stations). As we see, this circumstance did little to change the results.

The numbers N_Σ of earthquakes shown in the last column of Table 6, obtained by the complete analysis discussed above, were found to be much smaller than all previous determinations. The chief cause of this is found in the fact that during the present computations the following two circumstances were taken into account, which for some reason were not previously accounted for: 1) the presence of a systematic variation in magnitudes of explosions for the second zone (the zone of diffraction), which impelled us to exclude the second zone from calculations of average values; 2) the difference between local M_L and teleseismic (or unified) $\underline{m}$ magnitudes, established by Gutenberg previously for earthquakes [3] and confirmed by us for explosions.

Above we gave the geophysical basis for the necessity of considering these important circumstances. Only an analysis and computation of both circumstances has permitted us to eliminate those clear contradictions of a geophysical character that unavoidably arise when these circumstances are not kept in mind.

From a comparison of the second, third, and fifth columns of Table 6 it may be seen that, despite the position of the "Work Report" [10] that in the light of the new seismic data it is necessary to increase considerably the number N_Σ of earthquakes, a more attentive examination of these same new seismic data leads to the opposite conclusion. For explosions with yields of 1-20 kilotons, it is necessary to decrease the earlier Geneva evaluation of 1958 of the mean for the first and third zones to approximately one-half or one-third. For the first zone separately, it is necessary to diminish the number yet more: by a factor of 2.5-4. For this zone the number $N_\Sigma = 1200$ earthquakes for an explosion of yield Y = 5 kilotons, according to new, more reliable evaluation, proves to be less than the number (1500) previously ascribed to an explosion of 20 kilotons. We have distinguished the first zone because it is precisely from observations in this local zone that we must judge the effectiveness of the criterion of direction of first motion for distinguishing explosions (the use of this criterion is generally excluded in the second zone). As is known, this criterion was the only one used by all the delegations at both conferences in Geneva, in 1958 and 1959. In the future it is obvious that new criteria will be developed, more effective for the first as well as for the other zones.

Figure 10 shows graphs of the relationship between number of earthquakes N_Σ and the yield of an explosion, corresponding to the second, third, and fifth columns of Table 6. It is clear that the slope of our straight lines $N_\Sigma(Y)$ (the middle one for the first and third zones taken together, and the two lateral ones for each of these zones separately) is near the slope of the line drawn from the earlier Geneva data of 1958. The line of the "Work Report" of January 5, 1959, is distinguished from all the other lines by a distinctly greater slope and by a higher level throughout the principal range of yields for explosions: from 1 to 20 kilotons.

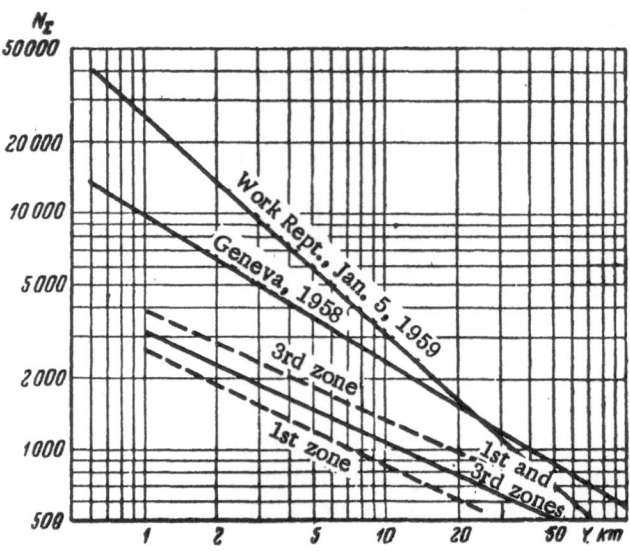

Fig. 10. Annual number N_Σ of shallow-focus earthquakes in the earth exceeding in magnitude $\underline{m}$ underground nuclear explosions of a given yield Y.

Let us remember that all the figures cited here concern the number of earthquakes for the entire world. The number of continental earthquakes is approximately half this number.

We should still emphasize that all the determinations we have made of the seismic effect of underground nuclear explosions refer, of course, to the conditions actually obtaining at the American nuclear test site in Nevada. However, the methods here used for analyzing and interpreting the magnitudes of concentrated underground explosions (either nuclear or chemical), and also the marked peculiarities in comparing the seismic effects of explosions and of earthquakes, apparently hold, in general outlines, for other continental areas.

Summary

The analysis in the present paper of American and Soviet observations on the seismic effect, and, specifically, on the magnitudes of the American underground explosions Rainier, Logan, and Blanca, with yields on the order of 1-20 kilotons, and also of other explosions of the Hardtack II series, of much smaller yields, has shown that, despite the statements in the American "Work Report" of January 5, 1959 concerning "New Seismic Data," the earlier Geneva determinations of 1958 on the number of earthquakes corresponding in magnitude to the explosions were not too low.

Moreover, a careful analysis of the new seismic data has led to the conclusion that the numbers of earthquakes are, on the average, but one-half to one-third the values obtained from the earlier Geneva evaluations for explosions in the given range of yields.

From this point of view, the task of any future international organization for control of underground nuclear explosions and for distinguishing explosions from earthquakes is not made more difficult, but rather made easier.

The author expresses his heartfelt thanks to his fellow workers, the Soviet experts at the conference in Geneva in 1959, for their friendly support, and also to his American colleagues, Doctor Carl Romney and Professors Hans Bethe, Frank Press, and John Tukey for their ardent discussions, which aided considerably in allowing us to make detailed analysis of this interesting and important matter and to distinguish what was essential from what was secondary.

LITERATURE CITED

1. C. F. Richter, "An instrumental magnitude scale," Bull. Seism. Soc. Amer. 25 (1935).
2. B. Gutenberg and C. F. Richter, "Magnitude and energy of earthquakes," Ann. Geofis., Roma 9, No. 1 (1956).
3. B. Gutenberg, "The energy of earthquakes," Quart. Journ. Geol. Soc., London, 112, 445 (1956).
4. C. F. Richter, Elementary Seismology (Freeman Co., San Francisco, 1958).

5.   Yu. V. Riznechenko, "Study of the seismic process," Izv. AN SSSR, seriya geofiz. No. 9 (1958).

6.   V. I. Buné, M. V. Gzovskii, K. K. Zapol'skii, V. V. Keilis-Borok, V. N. Krestnikov, L. N. Malinovskaya, I.L. Nersesov, G. I. Pavlova, T. G. Rautian, G. I. Reisner, Yu. V. Riznichenko, and V. I. Khalturin, "Methods for detailed study of seismicity," Trudy Inst. fiziki zemli AN SSSR, No. 9 (176) (1960).

7.   L. F. Bailey and C. F. Romney, "Seismic waves from the Nevada underground explosion of September 19, 1957, Bull. Geol. Soc. Amer. $\underline{69}$, 1672 (1958).

8.   Atomic Energy Commission Release on Hardtack Bomb Tests, No. 2-39, March, 1959.

9.   C. F. Romney, "Amplitudes of seismic body waves from underground nuclear explosions, Journ. Geophys. Res. $\underline{64}$, No. 10 (1959).

10. Work Report on the Question of Seismic Data Presented by the Delegation of the United States on January 5, 1959 [Russian translation] at the Conference for Banning Tests of Nuclear Devices.   Data of the Geneva Conference of Experts, 1959.

11. S. L. Solov'ev, "Some results of using the scale of earthquake intensities at seismic stations of the USSR." Studia geoph. et geod., 2, Praha (1958).

12. D. S. Carder and W. K. Cloud, "Surface motion from large underground explosions, Journ. Geophys. Res. $\underline{64}$, No. 10 (1959).

13. B. Gutenberg and C. F. Richter, "Earthquakes' magnitude, intensity, energy, and acceleration (Second Paper) Bull. Seism. Soc. Amer. $\underline{46}$, No. 2 (1956).

14. A. L. Latter, E. A. Martinelli, and E. Teller, "A seismic scaling law for underground explosions," Physics of Fluids (1959).

15. J. V. Riznichenko, "On quantitative determination and mapping of seismic activity," Ann. Geofis. Roma $\underline{12}$, No. 2 (1959).

16. I. L. Nersesov and Yu. V. Riznichenko, "The recurrence of earthquakes and a map of seismic activity," in the Collection: Seismic and Glacial Investigations during the International Geophysical Year [in Russian]. Collection No. 2 (Izd. AN SSSR, Moscow, 1959).

17. Yu. V. Riznichenko and I. L. Nersesov, "Development of a basis for a quantitative method of regional seismic zoning." Byull. Soveta po seismologii AN SSSR, No. 8 (1960).

18. E. K. Fedorov, M. A. Sadovskii, Yu. V. Riznichenko, V. I. Keilis-Borok, I. P. Pasechnik, A. I. Ustyumenko, and K. E. Gubkin, "An agreement to stop nuclear testing must be reached," Pravda, February 8, 1960.

19. I. P. Pasechnik, S. D. Kogan, D. D. Sultanov, and V. I. Tsibul'skii, "Results of seismic observations on underground nuclear and TNT explosion." Present volume.

20. V. I. Keilis-Borok,"Differences in the spectra of surface waves from earthquakes and from underground explosions." Present volume.

DIFFERENCES IN THE SPECTRA OF SURFACE WAVES
FROM EARTHQUAKES AND FROM EXPLOSIONS

V. I. Keilis-Borok

This paper investigates the relationship between spectra of surface waves in layered media and the dimensions and form of the source.

A distinctive feature of underground explosions is defined: the predominant period of surface waves from explosions is no more than one-fourth the period typical of surface waves from earthquakes of approximately the same intensity and epicentral distance; the propagation path for waves of comparable sources should be approximately the same.

Introduction

The frequency spectrum of interference surface waves of the Rayleigh and Love type in layered media depends on three basic factors: the frequency characteristics of the medium, the frequency spectrum of the forces or directions at the sources, and the spatial distribution of intensity at the source. The first two factors have an obvious physical significance and have been repeatedly investigated. The third factor has not yet been studied to any great extent, although its influence happens to be substantial [2, 4] and is relatively simple to compute. The present paper investigates the relationship between the spectrum of surface waves and the spatial properties of the source in the simplest models of an explosion and an earthquake. It should be mentioned that the second and third factors are somewhat interrelated (for example, the "natural period" of a source increases with the dimensions of the source). We shall here consider the third factor, the effect of the form of a source on the spectrum of surface waves and the distribution of intensity at the source for a given frequency spectrum of excitation stresses (the latter may be practically defined by the spectrum of body waves, although not always directly).

1. Initial Formulas

1. We shall use a solution to the following problem, given in several papers [2, 5, and others]: given the half space $z > 0$, covered by a finite number of plane-parallel layers. The medium in the layers and the half space is homogeneous, isotropic, ideally elastic, with fixed contacts. A nonsteady source acts in the medium. The object is to find the asymptotic representation of displacements at great distances from the source, computed along the layers.

The principal part of the displacements is found in waves of the Rayleigh and Love type. They are described in detail in the literature [1-5, 9-12].

2. Let us first assume that a concentrated force is applied to the surface $z = z_0$ at point $r = 0$ (r is the distance from the z axis), the spectrum of the source (Fourier transformation) being equal to $AQ(p)$ at time t, where p is the frequency of rotation.

The spectrum of displacement in a given harmonic of Rayleigh or Love surface waves at large values of <u>r</u> then has the form [1, 2, 5, 10]

$$U = A P_U \left(\frac{p}{v}, z_a \right) Q(p) \frac{1}{\sqrt{r}} e^{-i \frac{p}{v} r}. \tag{1}$$

Here U is the spectrum of one component of displacement (vertical or horizontal radial for the Rayleigh wave, horizontal tangential for the Love wave), A is a constant depending on the intensity of the source, P_U is the frequency characteristic of the medium (at a given position of the source P_U depends also on the direction of the force: Rayleigh waves are excited by vertical and radial components, Love waves by tangential components of the force), and <u>v</u> is the phase velocity of the given harmonic.

Formulas for $v(p)$ and P_U and a number of computations may be found in the literature [1-5, 10, 12]. From equation (1) it is easy to discover that if the stresses at the source are distributed through some region Ω such that $A = A(r, \varphi)$, at large values of <u>r</u>

$$U \approx R_A \left(\frac{p}{v}, \varphi \right) P_U \left(\frac{p}{v}, z_0 \right) Q(p) \frac{1}{\sqrt{r}} e^{-i \frac{p}{v} r}, \tag{2}$$

where R_A is the spectrum of the source according to distance:

$$R_A = \int_\Omega A(r, \varphi) e^{i r_0 p / v} d\Omega, \tag{3}$$

r_0 being the distance from the point of observation to the point in the region Ω.

Thus, the spectrum of U(p) is proportional not to the intensity of the source as a whole, but only to the component R_A, the wave length λ of which coincides with $2\pi \frac{v}{p}$, the length of the surface wave. This conclusion differs somewhat from the arrived at in other papers [2, 4]; it is easy to generalize the conclusion for the case of spherical or cylindrical boundaries, nonideal elasticity, etc.).

3. We shall now consider two different sources I and II (axially symmetrical with a radius <u>d</u> and forming an elongated rectangle with the sides D and <u>d</u>), creating, on some horizontal plane, stresses with the following spectra (below, the spectra of the stresses σ_z, τ_{rz}, etc. are designated respectively by $\tilde{\sigma}_z, \tilde{\tau}_{rz}$, etc.):

$$\text{or} \qquad \text{I} \quad \begin{cases} \tilde{\sigma}_z = A_n(r) Q(p), & \tau_{rz} = \tau_{z\varphi} = 0 & (4) \\ \tilde{\tau}_{rz} = A_n(r) Q(p), & \tau_{r y} = \sigma_z = 0, & (5) \end{cases}$$

$$A_n = 0 \quad \text{at} \quad r \geqslant d, \quad A_n = (d^2 - r^2)^n \frac{n+1}{\pi d^{2(n+1)}} \quad \text{at} \quad r < d, \tag{6}$$

$$\text{or} \qquad \text{II} \quad \begin{cases} \tilde{\sigma}_z = B(x, y) Q(p), & \tau_{rz} = \tau_{r\varphi} = 0 & (7) \\ \tilde{\tau}_{zx} = B(x, y) Q(p), & \tau_{zy} = \sigma_z = 0, & (8) \end{cases}$$

where B(x,y) = 0 is outside the rectangle $|x| \leq D$, $|y| \leq d$; inside this rectangle

$$B(x, y) = \pm \frac{3\pi}{16d} (d^2 - y^2) \cos \frac{\pi}{2D} x. \tag{9}$$

The upper and lower signs correspond to positive and negative y; D > d. The constant factors in equations (6) and (9) were introduced to simplify computations and have been so expressed that the total intensity of the source does not depend on <u>d</u> or <u>n</u>; it is easy to see that for (9) it is proportional to D.

Different intensity distributions at the source correspond to different values of <u>n</u> in equation (6); the greater <u>n</u> is, the more smoothly the stresses at the boundary of the source return to zero, the (n + 1) derivative stress produces rupture.

It is not difficult to see that for source I

$$R_{A_n} = \frac{J_{n+1}\left(p\frac{d}{v}\right)}{\left(p\frac{d}{v}\right)^{n+1}} C_n \qquad (C_0 = 2;\ C_1 = 8;\ C_3 = 384), \tag{10}$$

if the vertical stresses (4) are assigned. For the radial stresses (5)

$$R_{A_0} = 2i\frac{S}{\left(\frac{pd}{v}\right)^2};\quad R_{A_1} = 4i\left[\frac{S}{\left(\frac{pd}{v}\right)^2}\left(1 + \frac{3v^2}{p^2d^2}\right) + \frac{J_0\left(\frac{pd}{v}\right)}{\frac{pd}{v}} - 3\frac{J_1\left(\frac{pd}{v}\right)}{\left(\frac{pd}{v}\right)^2}\right], \tag{11}$$

where J_k is the Bessel function of the first kind:

$$S = \int_0^{pd/v} J_1(x)\,x\,dx.$$

For the source II, when the vertical stresses (7) are acting, by assuming $\frac{pd}{v}\sin\varphi = e$, $\frac{pD}{v}\cos\varphi = L$, we obtain

$$R_B = \frac{D}{d}\frac{1.5i}{e^3}(e^2 - 2e\sin e + 2 - 2\cos e)\left(\frac{\cos L}{1 - \frac{4L^2}{\pi^2}}\right). \tag{12}$$

If we multiply the right side of this equation by $\cos\varphi$, we obtain R_B for the horizontal stresses (8).

Let us now consider the effect of a center of expansion. As T. B. Yanovskaya [10] has shown (see [2] as well), a source within a layer is equivalent to specially selected stresses and displacements at the boundaries of the layer. In particular, the center of expansion within the upper layer is equivalent to those stresses σ_z, τ_{rz} at the upper (free) boundary for which

$$R_{\widetilde{\sigma}_z} = \frac{Q(p)}{\lambda_1 + 2\mu_1}\frac{p^4\gamma\operatorname{sh}\alpha h}{\alpha}, \tag{13}$$

$$R_{\widetilde{\tau}_{rz}} = -\frac{2iQ(p)}{\lambda_1 + 2\mu_1}\frac{p^3\operatorname{ch}\alpha h}{v}, \tag{14}$$

where

$$\alpha = p\sqrt{\frac{1}{v^2} - \frac{1}{a_1^2}},\quad \gamma = \frac{2}{v^2} - \frac{1}{b_1^2};$$

λ_1 and μ_1, being the Lamé constants in the upper layer, a_1 and b_1 the velocities of longitudinal and transverse waves in the upper layer, and $\underline{h}$ the depth to the source.

4. Above we have considered the spectrum of displacement, i.e., essentially a solution of a problem involving steady conditions.

We may transfer this directly to nonsteady displacement by means of the following formulas.

Let us designate $p = \bar{p}_m$ the root of the equation

$$r = tV(p), \tag{15}$$

where V is the group velocity. The value of $\underline{p}$ at which the function V(p) is at a maximum or at a point of inflection we shall designate by $p.^*$

If $\bar{p}_m$ is not too near one of the values of p^*, the nonsteady displacements $\underline{u}$ have the form [1, 2, 9-11]

$$u(t, r) = A\frac{1}{r}\sum_m U(\bar{p}_m)\exp\left[i\bar{p}_m\left(t - \frac{r}{v(\bar{p}_m)}\right)\right]\left[\frac{\partial}{\partial p}V(p)\right]_{p=\bar{p}_m}^{-\frac{1}{2}}\sqrt{2\pi i}. \tag{16}$$

If $\bar{p}_m$ is near p^*, such that $\left.\dfrac{\partial^2 V(p)}{\partial p^2}\right|_{p=p^*} \neq 0$, the corresponding item in equation (16) should be replaced by the expression

$$u(t,r) = A\sqrt{\pi}\,\frac{1}{r^{-\frac{5}{6}}}\,U(p^*)\exp\left[ip^*\left(t-\frac{r}{v(p^*)}\right)K(p^*)\,a\left[\left(t-\frac{r}{v(p^*)}\right)K(p^*)\frac{1}{r^{-\frac{1}{3}}}\right],$$

where $\underline{a}$ is the Airy function,

$$K(p) = \left[\frac{\partial^2 V(p)}{\partial p^2}\right]^{-1/3}[2V^2(p)]^{1/3}.$$

(17)

Mote detailed studies of non-steady oscillations are discussed in [1, 2, 9-11].

5. We shall now examine source II as a simplified model of an earthquake focus. At this source we shall not consider diffraction at the discontinuity, that stresses should be applied not on a horizontal but on an inclined or vertical surface, or related factors. However, for our task— to study the effect of elongation of the source— this simplification is not significant, especially since we shall be interested not in the spectrum itself of the surface waves but in any change in this spectrum. The superposition of sources (4) and (5) makes a model of a near-surface explosion [7]. The center of expansion (13) and (14) makes another, essentially equivalent, model of an explosion.

Waves from an explosion with a spherical cavity of radius R, the pressure on the walls of which is F(t), have been considered in [15], with due respect to the marginal conditions at the boundary of the cavity.

In order to compute the effect of these conditions it is sufficient to use a value of Q(p) in (13) and (14) corresponding to the solution obtained in [15].

If F(t) = 0 at t < 0 and F(t) = A at t > 0, the appropriate value of Q(p), as may be easily shown, has the form

$$Q(p) = \frac{AR^3}{6p_0^2\pi\sqrt{2}}\,\frac{\sqrt{2}+ip/p_0}{(1/\sqrt{2}+ip/p_0)^2+1},$$

where $p_0 = \dfrac{2\sqrt{2}a_0}{3R}$; and a_0 is the velocity of the longitudinal wave.

In a similar way we may investigate the spectrum of aerial explosions, when the immediate source of the stresses at the boundary is an aerial wave traveling at the velocity $\underline{c}$.

It is easy to see that, in this case,

$$Q(p)R_A = \int_0^\infty \frac{P(p\sqrt{r^2+h^2})}{R(r^2+h^2)}\,e^{-ip\frac{\sqrt{r^2+h^2}}{c}}\,J_\nu\left(\frac{p}{v}r\right)r\,dr.$$

Here $\underline{h}$ is the height of the explosion, R defines the attenuation of the aerial wave with distance, $\underline{p}$ is the spectrum of pressure in the aerial wave according to time, and $\nu = 0$ or $\nu = 1$ depending on whether we are considering the vertical or horizontal pressure.

It appears that, practically, $R = r^2 + h^2$; p, as a first approximation, may be considered not to depend on $\underline{r}$, and it is sufficient to investigate the vertical pressure. The inverse effect of surface waves on air is easily computed by means of the apparatus described in [1, 2, 12], but it seems hardly essential.

2. Some Calculations

1. The functions R_{A_n} and R_B occur as factors in the expression for the spectrum of surface waves and they define the effect of the form of the source and the distribution of intensity in the source.

From the determination of these functions (3) and from certain properties of the Fourier transformation we may already draw a number of practical conclusions.

Thus, it is interesting to note the possibility of regulating the intensity of the high-frequency part of the spectrum by changing the distribution of stresses at the source (this intensity becomes greater the more sharply the stresses change along the ordinates).

Figures 1 and 2 show graphs of R_{A_n} for axially symmetrical stresses, for vertical (4) and radial (5) stresses respectively. In the upper right of Fig. 1, graphs of A_n (distribution of intensity along the radius) are given for increased clarity.

Figure 3 shows graphs of R_B, indicating the effect of the form of elongated source II on the spectrum.

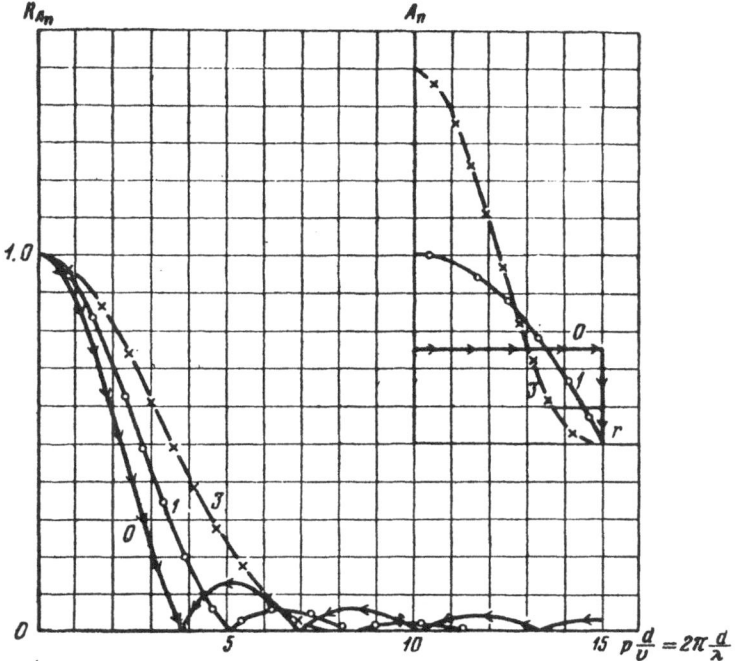

Fig. 1. Graphs of R_{A_n} for axially symmetrical vertical stresses (4). d) Radius of source, λ) wave length; the figures on the curve refer to values of $\underline{n}$. In the upper right— distribution of intensity at the source, corresponding to different values of $\underline{n}$.

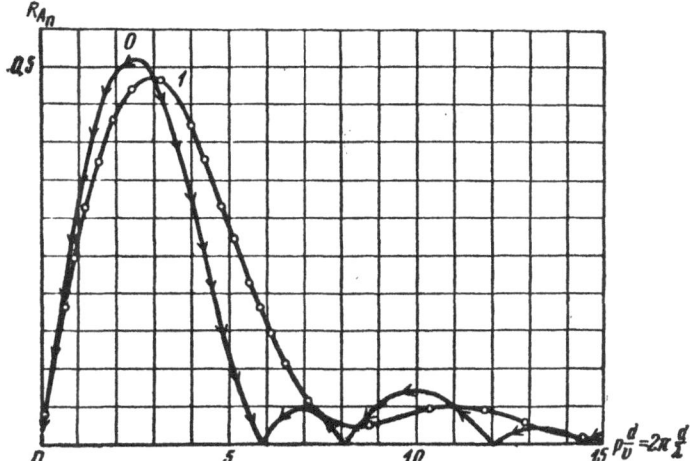

Fig. 2. Graphs of R_{A_n} for axially symmetrical vertical stresses (5). d) Radius of source, λ) wave length; figures on the curves refer to values of $\underline{n}$.

Let us formulate some conclusions:

a) As seen from Figs. 1 and 2, the resonant wave length is proportional to d [this circumstance should not be confused with the increase in natural period that accompanies increase in size of explosion cavity [15], which may occur for Q(p)].

b) For radial stresses the resonance wave length is somewhat less than for vertical stresses.

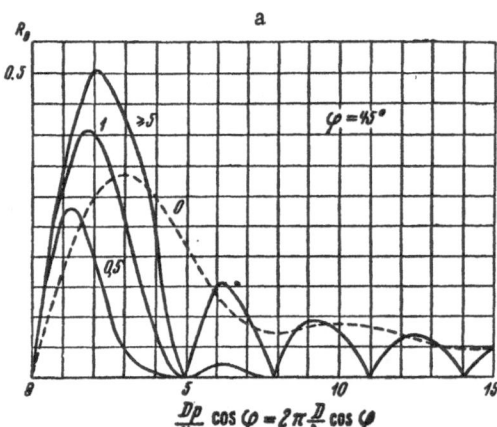

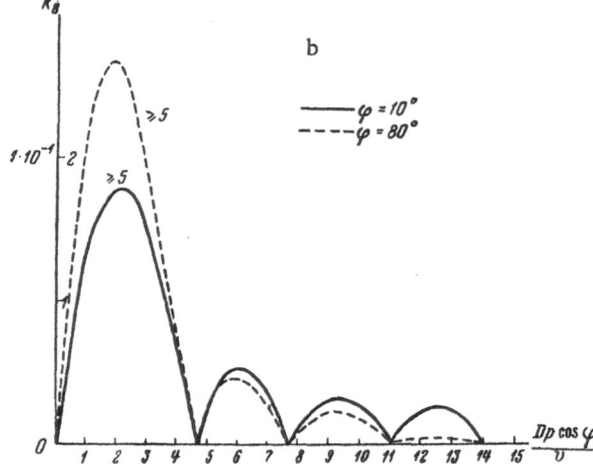

Fig. 3. Graphs of R_B for the elongated source II for applied vertical stresses (7). For the case of horizontal stresses (8), the ordinates of the curves must be multiplied by $\cos\varphi$. The figures on curves refer to values of D/d; a) $\cos\varphi$ ($\varphi = 45°$); b) $\cos\varphi$ ($\varphi = 10°$) and $\cos\varphi$ ($\varphi = 80°$).

c) The decrease in intensity with diminution of wave length is markedly irregular; in this, the more sharply the vertical stresses are reduced to zero at the boundary, the weaker is the attenuation of long waves (above 1.5-2) and the stronger is the attenuation of short waves (Fig. 1). At a rather sharp boundary of the source, the oscillation intensity drops sharply—tenfold—at $\lambda = d$, but with further decrease in λ the change is comparatively gradual and uneven.

d) For the elongated source II— the model of a focus — the resonance wavelength is on the order of 2 D $\cos\varphi$, i. e., a section of the source along a straight line connecting its center with the point of observation.

Consequently, the resonance wave length depends on the azimuth and is a maximum for small values of $\cos\varphi$ on the extension of the axial line of the focus.*

Let us now see what effect the frequency characteristics of the medium P_U have on these conclusions.

According to (2) the product RP_U is the complete frequency characteristic of the system medium—source; the product of this characteristic and the spectrum of excitation Q(P) defines the final spectrum of oscillations in the surface wave. Since P_U depends on the ratio H/λ (H being the thickness of the layers) and R on D/λ (D being the linear extent of the source), the complete characteristic of RP_U depends also on the ratio D/H.

Computations of P_U are rather cumbersome and they are made only for some apparently typical cases.

We shall use the computations of P_U from [2, 10] for the vertical component of displacement of the Rayleigh wave in a layer on the half space; the following ratios of constants obtain:

$$a_2/a_1 = 1.4; \quad a_2/b_2 = 1.8; \quad a_2/b_1 = 2.3; \quad \mu_2/\mu_1 = 1.9$$

(a and b being the velocity of longitudinal and transverse waves, μ the shear modulus, and the indices 1 and 2 refer to the layer and to the half space).

Figures 4-6 show the products RP_U for the sources I and II. It is easy to see that the resonance properties of the layer in the examined case do not mask the difference between R_{A_n} and R_B.

---

* This opens up the fundamental possibility of using the spectrum of surface waves for studying the mechanism of earthquakes. A practical difficulty is the exclusion of effects that differences in structure of the medium have on the paths of waves traveling in different directions.

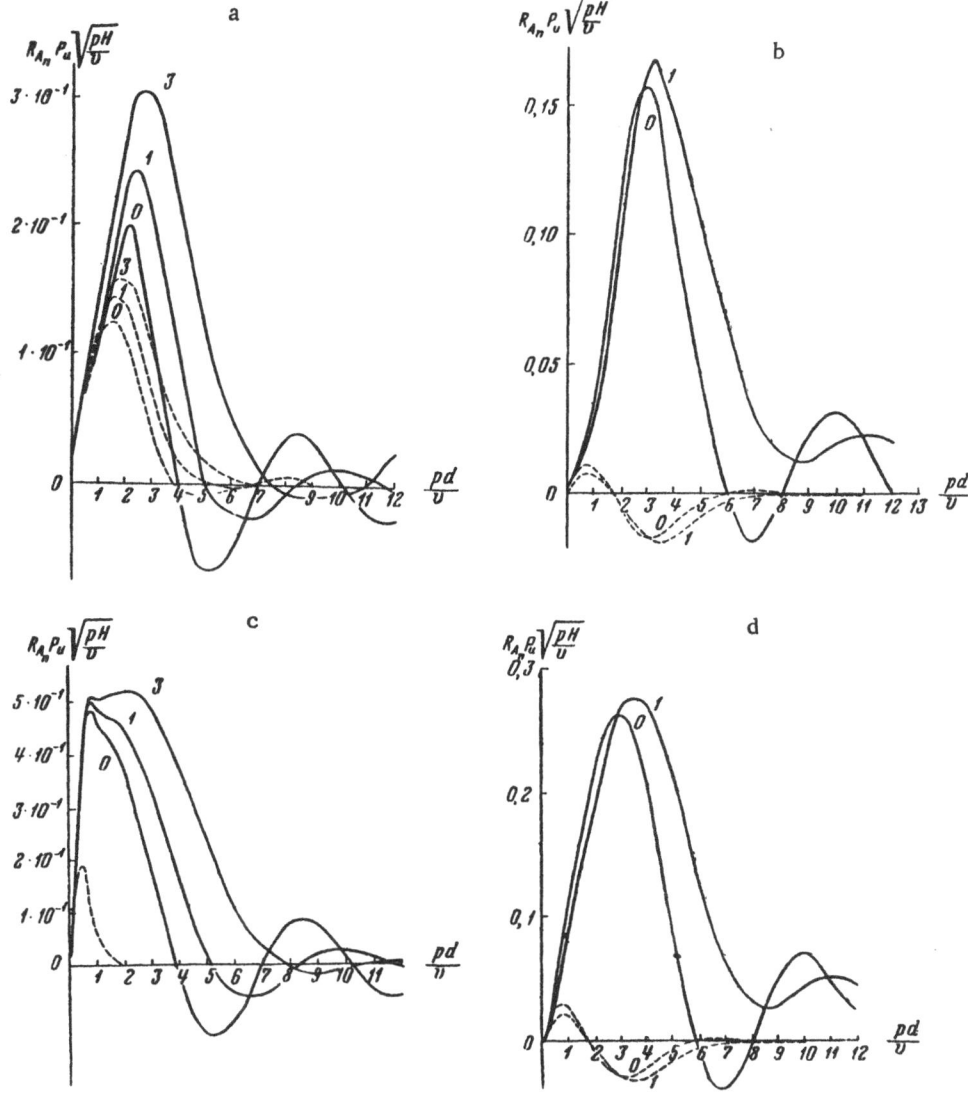

Fig. 4. Frequency characteristics of the system layer—source for the axially symmetrical source I. H) Thickness of layer; h) depth of source; the figures on the curves refer to values of n; k = d/H; the solid line is for h = 0, the dashed line for h = H; a) vertical stresses (k = 1); b) vertical stresses (k = 0.2); c) horizontal stresses (k = 1); d) horizontal stresses (k = 0.2).

The effect of the frequency characteristic of the medium is manifested chiefly in reduction of wave lengths; the thicker the layer the more marked this reduction, a fact that has obvious physical significance. The change in resonance wave length with azimuth (approximately as $\cos \varphi$) is preserved with due consideration of the layer.

3. Differences in the Spectrum of Surface Waves for the Adopted Models of an Earthquake and an Explosion

1. Formula (2) and the computations in section 2 permit us to compare the theoretical spectrum of explosions and earthquakes. Since the frequency characteristic of the medium P_U does not depend on the mechanism of excitation, it may be taken, for the time being, to be the same for both sources. The spectrum of excitation $Q(p)$, roughly speaking, is characterized by the spectrum of body waves. It is known that somewhat higher-frequency body waves are excited during explosions; consequently, in the worst case, it may be assumed that $Q(p)$ is also the same for both

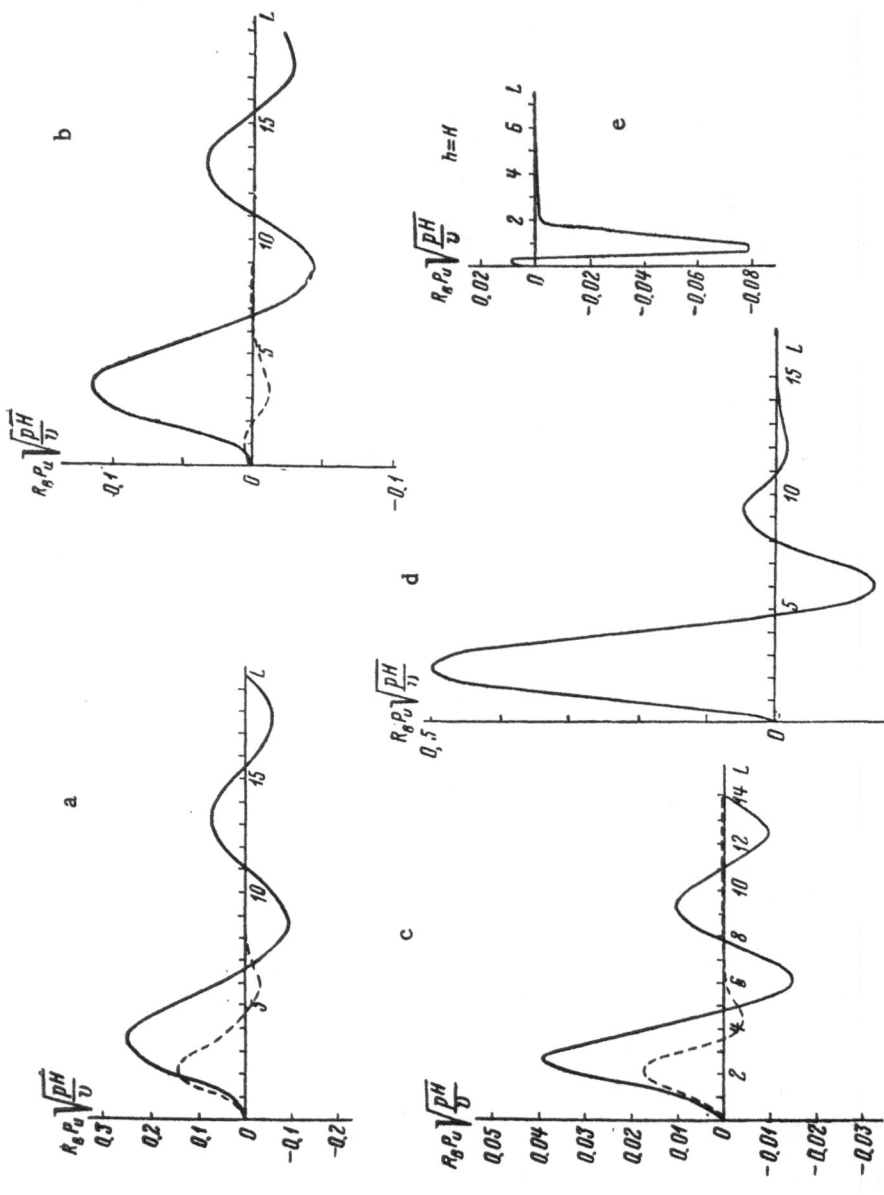

Fig. 5. Frequency characteristics of the system layer—source for the elongated source II. H=Thickness of layer;

h=depth of source. $L = \dfrac{pD\cos\varphi}{v} = 2\pi\dfrac{D\cos\varphi}{\lambda}$, $k = \dfrac{D}{H}$. a) $\varphi = 45°$, $k = 1$; b) $\varphi = 45°$, $k = 1$; c) $\varphi = 10°$, $k = 1$;

d) $\varphi = 80°$, $k = 1$; e) $\varphi = 80°$, $k = 1$; solid line is for h = 0, dashed line for h = H.

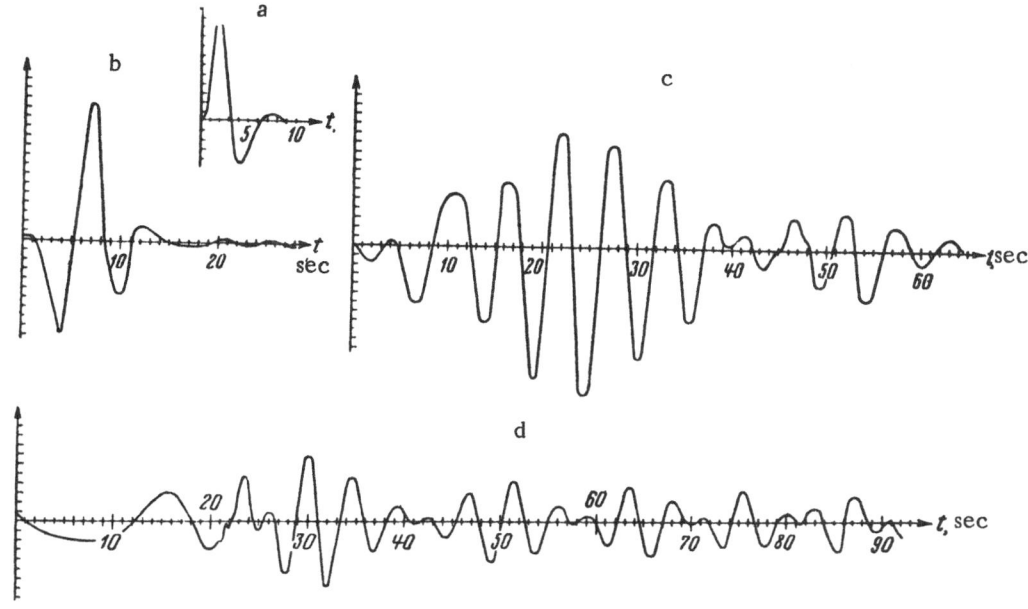

Fig. 6. Approximate theoretical seismogram of nonsteady Rayleigh waves. a) Form of initial impulse; b, c, and d) displaced to distances r, equal to 5H, 15H, and 45H, respectively. The vertical scales are not the same (the value of a division on Fig. 6b being approximately 2 and 2.5 times those on Figs. 6c and 6d).

sources. Then any differences in the spectra of surface waves will be defined by the R functions. These functions do not depend on the structure of the medium, and we may therefore obtain very general conclusions.

The computations in section 2 show that any difference between sources is defined principally by their dimensions. The maximum of the function R for the most variable forms of sources and distributions of intensities corresponds approximately to the same ratio of wave length to linear dimension of the source in the section including the epicenter and the station. In other words, the greater the size of the source, the greater the prevailing period of the surface wave. At the same time, the dimensions of an earthquake focus are considerably greater than the dimensions of the zone of an explosion (of equal intensity). Consequently, the periods of surface waves from earthquakes are considerably longer than those from explosions. An exception may be found in the case of small values of $\cos \varphi$, which will be considered below.

2. Let us make some quantitative evaluations. For earthquakes with energies E of the seismic waves we get

$$E \approx 2 \frac{\tau^2}{\mu} d^2 DK, \tag{18}$$

$$E \approx \frac{\tau^2}{\mu} d^3 L, \tag{19}$$

where τ represents average stresses, μ is the shear modulus, and K and L are coefficients having values near 1.

Formula (18) is the result of solving a two-dimensional problem; it corresponds to a strongly elongated focus of length 2 D and width 2 d; for displacement along the strike such a formula was obtained by Knopoff [16], and for displacement across the strike by Starr [17].

Formula (19) was obtained from [6] for a three-dimensional model of a focus, a circular explosion surface of radius d. If we insert D (the area of the surface of the explosion) in place of d, both formulas give approximately identical results, agreeing with experiment.

For typical values, such as $\tau \approx (-10) \cdot 10^7$ CGS, $\mu \approx 5 \cdot 10^{11}$ CGS, and setting L = 1, from (19) we obtain

$$d \approx (1 - 3) \, 10^{-1} \sqrt[3]{E}.$$

At the same time, as follows from M. A. Sadovskiĭ's formula, we get for explosions

$$d \approx 10^{-2} \sqrt[3]{\bar{E}}.$$

Approximately the same evaluation, perhaps even more favorable for distinguishing explosions, follows from data on the radius of the explosion cavities formed during the underground nuclear explosions of the Hardtack II series [14].

Consequently, the maximum of the function R for explosions fits a frequency approximately 3-10 times that for earthquakes of the same energy. For strong earthquakes, at D ≫ d, this difference is yet greater. These evaluations are, of course, approximate; the figures given may vary, the model of the sources may be made more complex, and so forth, but the general conclusions are definite.

3. At first glance it seems strange that the spectra of surface waves differ so strongly, whereas the spectra of body waves do not. This may be explained in the following way: during earthquakes the concentration of stresses is much lower than during explosions of the same energy, since the energy of an earthquake is spread through a greater volume than that surrounding the plane of an explosion. From this it follows that during explosions there will be a much larger zone of nonelastic deformation; but high frequencies are absorbed in this zone, and this may partly level out the spectrum of body waves [the function Q(p)]. The surface waves are formed outside the immediate vicinity of the focus, and plastic deformation affects them only indirectly, through the function Q(p).

The examined difference in the spectrum of surface waves should be magnified yet more by two other factors in equation (2). First, the depth to the source h, for earthquakes, is generally greater than for explosions, and with an increase in h the high frequencies begin to attenuate ever more quickly; this has a simple physical meaning (the high frequencies correspond to Rayleigh waves in the half space) and is illustrated by the computations in section 2. Secondly, the maximum of the function Q(p) for explosions always falls at somewhat higher frequencies.

On the other hand, the examined difference will be smoothed out by high absorption of the high frequencies (for example, during the nuclear blasts of the Hardtack II series, the absorption of high frequencies was such that the energy of the seismic waves decreased almost systematically throughout the first few kilometers).

4. Let us compare the theoretical conclusions we obtained with experimental data.

a) The difference in the spectra of surface waves from chemical explosions and from earthquakes was studied experimentally, and discussed in [7], for epicentral distances ranging from 250 to 2200 km. It was stated in [7] that during explosions the predominant period is approximately 2 sec, and during earthquakes the period increases with distance, from 5 to 10 sec.

These data are in agreement with the theory discussed. A possible explanation of why the difference in periods is smaller at short distances will be given below.

b) For an explosion of one kiloton of explosive material (communication of D. A. Kharin), the energy of the seismic waves amounted to $4 \cdot 10^{17}$ ergs, and the diameter of the shatter zone was 150-200 m. It is easy to compute from (14) that for an earthquake of the same energy, the area of the plane of rupture is 1 km across, the same as the length of the focus (on the order of a kilometer), and this is five or six times the long dimension of the shatter zone of the explosion.

The predominant period of surface waves from explosions is 2 sec; from earthquakes of the same intensity it is 8 sec, four times as great; this coincides with theoretical evaluations better than might be expected.

Frank Press has computed the spectra of surface waves using the numerical seismograms of the Blanca and Logan shots and of three earthquakes (report at the conference of the technical group of experts at Geneva, 1959). He also came to the conclusion that explosions excite more short-period surface waves. A more detailed analysis of his data is rendered difficult because it is uncertain in which scale he has determined the magnitudes of the indicated earthquakes.

5. Experience with the preceding work permits us to indicate an approach to practical recognition of underground explosions by the spectrum of surface waves.

We shall certainly not resort here to any cumbersome, precise determination of the amplitude spectrum; it is possible to determine the amplitude directly by visual inspection of the record of the predominant frequency. In doing this, one should keep in mind that, according to (16) and (17), the surface waves consists of two phases: one, the preliminary, is represented by oscillations (16) with continuously changing period $\bar{p}_m$; the other, attaching to the Airy phase (17), is represented by pulses with a discrete set of periods p.* The difference between these phases is clearly seen in Fig. 6, where we show an approximate, theoretical seismogram of a nonsteady Rayleigh wave; this wave was computed for the initial impulse as one of the forms identified by Press experimentally from the numerical seismograms of Rayleigh waves from earthquakes.

The periods of the Airy phase depend only on the structure of the medium; for a given medium it is possible to define them by computing the group velocity or by observation (by the statistical method developed by Gutenberg [13]). The periods of the preliminary phase and of the Airy phase should be studied separately, especially since the Airy phase dies out somewhat more slowly.

The distinguishing feature of explosions here considered reduces to the following: in the Airy phase to a discrete change from one period p* to another, in the preliminary phase to a decrease in the predominant period. In practice it is possible to orient oneself by the proper features on the record.

It is now easy to explain the experimental data [7]. The spectrum of explosions is apparently more homogeneous, and the observed waves correspond to a single Airy phase; for earthquakes the resonance properties of the layer corresponding to this phase lead only to diminished periods at close distances. In any case, at distances exceeding 1000 km, the experimental and theoretical data agree rather well; these distances are of interest only for the Geneva control system.

6. Let us determine the probability of doubtful cases. If the value of $|\cos \varphi|$ is rather small, i.e., if the point of observation is nearly perpendicular to the strike of the focus, the wave length will correspond to the transverse section of the focus, and the examined effect will be masked. Let us now take as the value of $|\cos \varphi|$ a distinctly higher value $|\cos \varphi| \leq \frac{1}{3}$.

The probability of this case is about 30% for a single station and is negligibly small for several stations if they do not lie near a single great-circle arc intersecting the epicenter.

Summary

The results here discussed, together with the experimental data [7], lead one to formulate the following distinguishing feature of underground explosions: the predominant period of the Rayleigh wave is no more than one-fourth the period typical of Rayleigh waves from earthquakes arising in the same epicentral zone and having the same magnitude and epicentral distance.

This feature may be used when there are distinct records of Rayleigh waves from at least three stations not lying on a single great-circle arc passing through the epicenter.

LITERATURE CITED

1. L. M. Brekhovskikh, Waves in Layered Media [in Russian] (Moscow-Leningrad, 1957).
2. V. I. Keilis-Borok, Interference Surface Waves [in Russian] (Moscow, 1960).
3. V. I. Keilis-Borok, "Surface waves in a layer lying on a half space," Izv. AN SSSR, seriya geofiz. (1951).
4. V. I. Keilis-Borok, "Interference waves in a multilayered medium," Doklady Akad. Nauk SSSR 95, No. 4 (1954).
5. V. I. Keilis-Borok, "Asymmetrical interference waves in a layered medium," Doklady Akad. Nauk SSSR 107, No. 4 (1956).
6. V. I. Keilis-Borok, "Estimation of the displacement in an earthquake focus and source dimensions," Ann. Geofis., Roma 12, No. 2 (1959).
7. S. D. Kogan, I. P. Pasechnik, and D. D. Sultanov, "The difference in periods of seismic waves arising from underground explosions and from earthquakes," Doklady Akad. Nauk SSSR 29, No. 6 (1959).
8. G. I. Petrashen', Data on the Quantitative Study of the Dynamics of Seismic Waves [in Russian], pt. 2 (Moscow-Leningrad, 1957).
9. K. Pikeris, "Theory of propagation of sound from an explosion in shallow water," from the Collection: Propagation of Sound in the Ocean [Russian translation], Izd. inostr. lit., (1951).

10. T. B. Yanovskaya, "Determination of the dynamic parameters of an earthquake focus from the records of surface waves," Izv. AN SSSR, seriya geofiz. No. 3 (1958).

11. T. B. Yanovskaya, "The problem of investigating dispersed surface waves near the minimum group velocity," Izv. AN SSSR, seriya geofiz. No. 12 (1959).

12. M. Ewing, F. Press, and Jardezki, "Elastic waves in a layered media (N. Y., 1957).

13. B. Gutenberg, U. S. Coast and Geodetic Survey, Spec. Publ. No. 201 (1936).

14. G. W. Johnson, G. H. Higgins, and C. E. Violet, Journ. Geophys. Res. 64, No. 10 (1959).

15. Sharp, Geophysics, No. 2 (1942).

16. L. Knopoff, Geophys. Journ. No. 1 (1958).

17. Starr, Proc. Camb. Phil. Soc. 24 (1928).

THEORETICAL MODEL OF AN EXPLOSION AT AN INTERFACE

T. I. Vavilova and B. Ya. Gel'chinskii

This paper investigates the directivity function for a center of expansion in the vicinity of an interface. It is shown that the first arrival of the longitudinal wave is directed away from the source (compressional wave), as in a homogeneous medium. Thus, the criterion of the first arrival remains in force for an explosion, near the interface.

1. The widely accepted mathematical model of an explosion is a center of expansion. It is easy to show that in a continuous nonhomogeneous medium the first movement of the longitudinal wave arising through the action of this source is compressional, i.e., in the direction of expansion; in the discussion to follow this direction of displacement will be considered positive.

In this paper it is shown that the sign of the first arrival remains positive even when the center of expansion is located near or on a discontinuity in the elastic constants, and the waves arriving first are much more intense than those following a short interval behind. For proof of this it is sufficient to examine the boundary between two homogeneous media.

2. The considered source may be obtained as the limit of superposition of two centers of expansion situated on opposite sides of the interface and corresponding to equal pressure on the surface of an elementary sphere. The field $\mathbf{u}$ of the source situated at the interface will be

$$\mathbf{u} = \frac{1}{2}(\mathbf{u}^+ + \mathbf{u}^-), \tag{1}$$

where $\mathbf{u}^+$ and $\mathbf{u}^-$ are the limiting values of the field excited by the center of expansion situated in the media $\underline{r}$ and q, respectively (Fig. 1), at the point $A^+(A^-)$ as it tends toward the boundary.

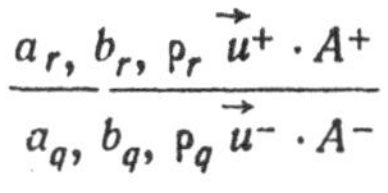

$$\frac{a_r,\ b_r,\ \rho_r\ \overrightarrow{u^+} \cdot A^+}{a_q,\ b_q,\ \rho_q\ \overrightarrow{u^-} \cdot A^-}$$

Fig. 1

On the basis of the general case of the ray method [2], an examination of the formulas and tables of refractive indices and a study of the formation of head waves [3], it may be asserted that the sign of the first movement of longitudinal waves arriving first at any point in the medium will be determined by the sign of the directivity function of the source. From this it follows that the direction of first movement of the fields $\mathbf{u}^+$ and $\mathbf{u}^-$, and consequently of $\mathbf{u}$ also, is in response to the compressional wave. Thus, the fact that the sign of the first arrival of the wave from the considered source is positive is shown to be trivial. But there is important practical significance in the intensity ratios of the different longitudinal waves generated in the given example.

Let us consider the wave picture corresponding to the fronts of the longitudinal waves (Fig. 2).

In Fig. 2 the symbol r is given to the medium in which the velocity of transverse waves v_{S_r} is least: $v_{S_r} < v_{S_q}$. It may be accepted that $v_{P_r} < v_{P_q}$. The critical angle is designated by $\tilde{\theta}_{P_r}$, so that

$$\sin \tilde{\theta}_{P_r} = \frac{v_{P_r}}{v_{P_q}}. \tag{2}$$

The segment of the front I corresponds to the direct longitudinal wave $\overset{\approx}{P}_q$, which formes by superposition of the fields of the direct wave P_q, the reflected waves P_qP_q, and the refracted wave P_rP_q. On the basis of equations of the ray method [2], the principal part of the field of this wave is shown to equal

$$u_I = \frac{\left\{ \delta^2 \sigma_0 [1 + (P_qP_q)] + \dfrac{m \cos \theta_{P_q}(P_rP_q)}{\cos \theta_{P_r}} \right\}}{8\pi \mu_r v_{P_q} R} K_{P_q} f(t - \tau_{P_q}), \tag{3}$$

where

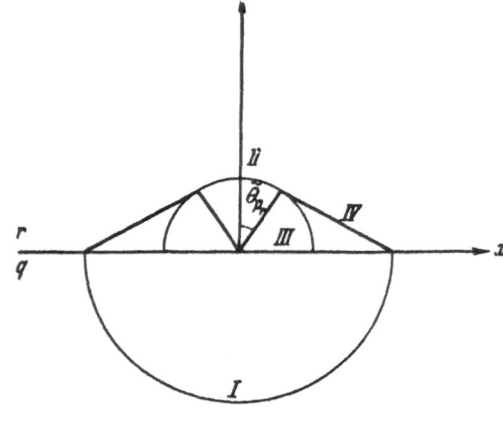

Fig. 2

$$\delta = \frac{v_{S_r}}{v_{S_q}}; \quad m = \frac{v_{P_r}}{v_{P_q}};$$

$$\gamma = \frac{v_{S_r}}{v_{P_r}}; \quad \Delta = \frac{v_{S_q}}{v_{P_q}}; \quad \sigma_0 = \frac{\rho_r}{\rho_q}; \tag{4}$$

$(P_qP_q) = |(P_qP_q)|e^{i\chi_{qq}}$ is the index of reflection from the table [3] (let us note at once that in the table in [3] it should read: $\chi_{qq} = m\pi$); K_{P_q} is a unit vector normal to the surface of the front, pointing in the direction of expansion of the wave P_q.

$$f(\Delta t) = \text{Im} \int_0^\infty |a(v)| e^{i[v\Delta t + \alpha(v)]} dv, \tag{5}$$

(f determines the form of the impulse in the vicinity of the front of the direct wave $\overset{\approx}{P}_q$); τ_{P_q} is the propagation time of the direct wave to the point of observation, and R is the distance along the ray from source to point of observation.

In the region of II, where $\theta_{P_r} < \tilde{\theta}_{P_r}$, we obtain for the wave $\tilde{P}_r$

$$u_{II} = \frac{\left[1 + (P_rP_r) + \dfrac{\delta^2 \sigma_0 \cos \theta_{P_r}}{m \cos \theta_{P_q}}(P_qP_r) \right]}{8\pi \mu_r v_{P_r} R} K_{P_r} f(t - \tau_{P_r}), \tag{6}$$

where

$$(P_rP_r) = |(P_rP_r)| \cos \chi_{rr}, \qquad (P_qP_r) = |(P_qP_r)| \cos \chi_{qr} \tag{7}$$

corresponding to the indices in the table.

The wave $\tilde{P}_r$ in segment III forms as a result of superposition of the following waves: direct P_r, P_rP_r (reflected at the critical angle), and P_qP_r (screened); in this segment the indices of reflection of P_rP_r and of refraction of P_qP_r will be complex, and this fact leads to a change in the form of the displacement. The main part of the field of the longitudinal wave in segment III is defined by the equation

$$u_{III} = \frac{|I_{III}|}{8\pi \mu_r v_{P_r} R} K_{P_r} f(t - \tau_{P_r}, \chi_{III}), \tag{8}$$

where

$$I_{III} = |I_{III}| e^{i\chi_{III}} = 1 + (P_rP_r) + \frac{\delta^2 \sigma_0 \cos \theta_{P_r}}{m \cos \theta_{P_q}}(P_qP_r),$$

$$f\left(\Delta t, \chi_{\text{III}}\right) = \text{Im} \int_0^\infty |a(\nu)| e^{i[\nu \Delta t + \alpha(\nu) + \chi_{\text{III}}]} d\nu. \tag{9}$$

In segment IV the head wave $\bar{P}_q \bar{P}_r$ will be recorded; it is easy to see that this is obtained from the head wave $P_r P_q P_r$ and the reflected wave $P_q P_r$ by the limiting transition $h \to 0$ and $\theta_{P_q} \to \pi/2$. In this the principal (zero) approximation for the field of the refracted wave tends toward zero [2], and therefore the principal part of the field of the head wave will be defined by the following approximation of the ray method, which, as is known, will be a lower-frequency approximation.

3. The computation of the first approximation of the wave $P_q P_r$ would require the tabulation of several cumbersome factors. In order to avoid this, we have used the principle of reciprocity [4] to obtain a formula for the field of the head wave $\bar{P}_q \bar{P}_r$.

The reciprocity principle for a source of the type grad $\delta(r)$ is formulated in the form [4]

$$\text{div } \mathbf{u}_{12}^{\bullet} = \text{div } \mathbf{u}_{21}^{\bullet}, \tag{10}$$

where $\mathbf{u}_{12}^{\bullet}(\mathbf{u}_{21}^{\bullet})$ represent the field of displacement at point 2 or 1, induced by the center of pressure grad $\delta(r)$, situated at point 1 or 2. Let point 1 coincide with point A, and point 2 be found at the front of the head wave. As has already been noted, we should consider the center of expansion corresponding to identical pressures at spheres in the media $\underline{r}$ and q (see Fig. 2).

On the basis of (1) and from the known relationship between the fields induced by these sources [1], we obtain the equations

$$\mathbf{u}_{12}^{\bullet} = \frac{1}{2} \left(\gamma^2 \mathbf{u}_{12}^+ + \Delta^2 \mathbf{u}_{12}^- \right), \qquad \mathbf{u}_{21}^{\bullet} = \frac{1}{2} \gamma^2 \left(\mathbf{u}_{21}^+ + \mathbf{u}_{21}^- \right), \tag{11}$$

where γ and Δ are defined by formulas in [4].

If we use the relations

$$\frac{\partial u_{21}^+}{\partial x} = \frac{\partial u_{21}^-}{\partial x}; \qquad \frac{\partial u_{21}^+}{\partial y} = \frac{\partial u_{21}^-}{\partial y}, \tag{12}$$

deriving from the discontinuity of displacements at a fixed contact between two media, then from (10) and (11) we obtain the equation

$$\text{div } \mathbf{u}_{12} = \frac{1}{2} \left(1 + \frac{\gamma^2}{\Delta^2} \right) \text{div } \mathbf{u}_{21}^+ + \frac{1}{2} \frac{\gamma^2}{\Delta^2} \left(\frac{\partial u_{21}^{-(z)}}{\partial z} - \frac{\partial u_{21}^{+(z)}}{\partial z} \right), \tag{13}$$

where $u_{21}^{(z)}$ is the projection of displacement on the $\underline{z}$ axis.

From this it is possible to determine the field of $\mathbf{u}_{12}$. The fields of $\mathbf{u}_{21}^+$ and $\mathbf{u}_{21}^-$ have the form

$$\mathbf{u}_{21}^+ = \mathbf{u}_{P_r P_q P_r} + \mathbf{u}_{P_r P_q S_r};$$
$$\mathbf{u}_{21}^- = \mathbf{u}_{P_r P_q} + \mathbf{u}_{P_r P_q S_q}. \tag{14}$$

From known formulas [2] we have the following relations for the principal parts of these fields:

$$\mathbf{u}_{P_r P_q P_r} = |J_{P_r P_q P_r}| K_{P_r} \int_0^t f(t - \tau_{P_r}) \, dt,$$

$$\mathbf{u}_{P_r P_q S_r} = |J_{P_r P_q S_r}| \cos \chi_{qr} \mathbf{t}_{S_r} \int_0^t f(t - \tau_{S_r}) \, dt,$$

$$\mathbf{u}_{P_r P_q} = \left[J_{P_r P_q}^{(0)} f(t - \tau_{P_q}) + J_{P_r P_q}^{(1)} \int_0^t f(t - \tau_{P_q}) \, dt \right] K_{P_q} + J_{P_r P_q}^{(2)} \int_0^t f(t - \tau_{P_q}) \, dt \, \mathbf{t}_{P_q},$$

$$u_{P_r P_q S_q} = J_{P_r P_q S_q} \int_0^t f(t - \tau_{S_q})\, dt\, \mathbf{t}_{S_q}, \tag{15}$$

where **t** is a unit vector tangent to the surface of the wave front, lying in the plane of incidence, and J is the intensity of the wave field.

If we use the relation

$$\frac{\partial \tau_e}{\partial z} = \frac{\cos \theta_e}{v_e}, \tag{16}$$

then from (13) and (15) and the known formulas for intensities of head waves we obtain the formula

$$u_{\widetilde{P}_q \widetilde{P}_r} = \frac{I_{\widetilde{P}_q \widetilde{P}_r}}{\sqrt{x l^3}}\, K_{P_r} \int_0^t f(t - \tau_{P_r})\, dt, \tag{17}$$

where

$$I_{\widetilde{P}_q \widetilde{P}_r} = \frac{\mathrm{tg}\,\widetilde{\theta}_{P_r}}{8\pi\mu_r}\left[1 + \frac{\gamma^2 m^2}{\Delta^2}\frac{\Gamma^{PP}(P)}{m} + \frac{\gamma^2}{\Delta^3}\cos\theta_{S_r}\Gamma^{PS}(P) + \frac{\gamma^5}{\Delta^3}\cos\theta_{S_q}\Gamma^P_S(P)\right]; \tag{18}$$

l being the distance along the boundary between point 1 and point 1' (Fig. 3), and $\underline{x}$ being the horizontal distance to point 2. $\Gamma^{PP}(P)$, $\Gamma^{PS}(P)$, and $\Gamma^P_S(P)$ are the reflective indices for the head waves $\widetilde{P}_r P_q \widetilde{P}_r$, $P_r P_q S_r$, and $P_r P_q S_q$. The meanings of the indices $\Gamma^{PP}(P)$ and $\Gamma^{PS}(P)$ are given in [3]. The index $\Gamma^P_S(P)$ was computed specially and proved to be very small.

4. The final computed formulas have the following forms.

Wave $\widetilde{P}_q$ (Zone I)

$$u_{\mathrm{I}}8\pi\mu_r v_{P_q}R = \left\{\delta^2\sigma_0\left[1 + (P_q P_q)\right] + \frac{m\cos\theta_{P_q}}{\cos\theta_{P_r}}(P_r P_q)\right\}f(t - \tau).$$

Wave $\widetilde{P}_r$ (Zone II)

$$u_{\mathrm{II}}8\pi\mu_r v_{P_r}R = \left[1 + (P_r P_r) + \frac{\delta_2}{m}\sigma_0\frac{\cos\theta_{P_r}}{\cos\theta_{P_q}}(P_q P_r)\right]f(t - \tau).$$

Wave $\widetilde{P}_r$ (Zone III)

$$u_{\mathrm{III}}8\pi\mu_r v_{P_r}R = I_{\mathrm{III}}f(t - \tau, \chi_{\mathrm{III}}).$$

Wave $\widetilde{P}_q\widetilde{P}_r$ (Zone IV)

$$u_{\mathrm{IV}}8\pi\mu_r v_{P_r}R = 8\pi\mu_r I_{\widetilde{P}_q\widetilde{P}_r}M\left[\int_0^t f(t - \tau)\, dt\right],$$

where $\quad M = \dfrac{Tv_{P_r}R}{2\pi\sqrt{x l^3}};$

$$\frac{T}{2\pi}\left[\int_0^t f(t - \tau)\, dt\right] = \int_0^t f(t - \tau)\, dt.$$

The scale factor M is taken as unity. There is practical interest in the case when the length of the path l along the boundary is small in comparison with the path R from the boundary. In these cases M is naturally greater than unity, and the comparative intensities of the head waves are low in in the computations.

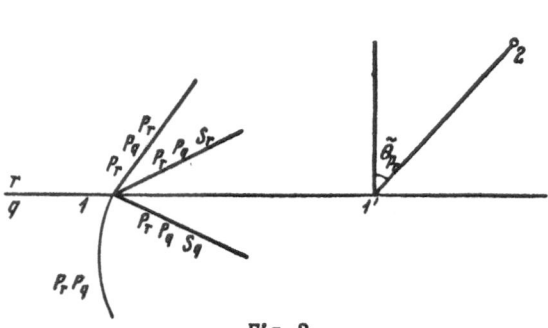

Fig. 3

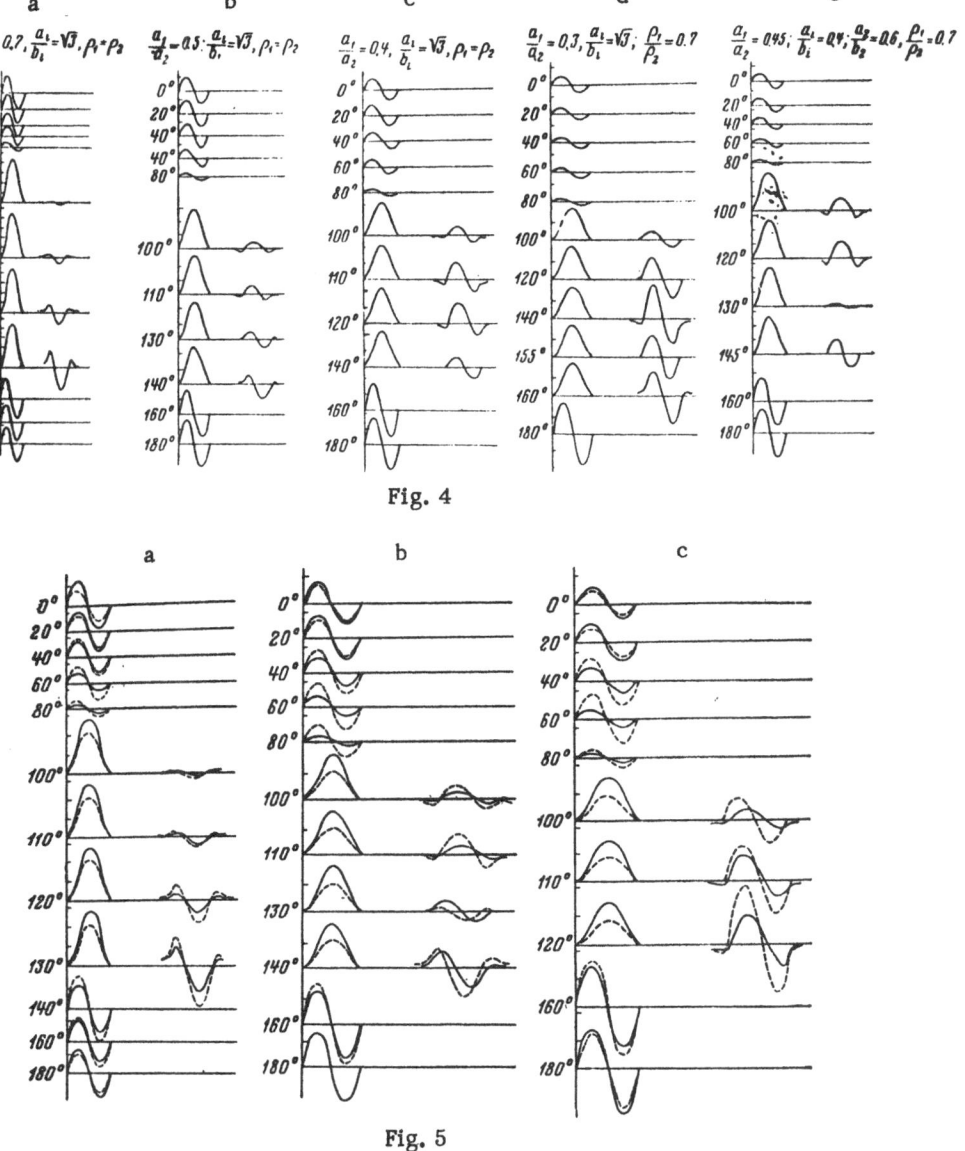

Fig. 4

Fig. 5

5. Figure 4 shows the results of the computation: theoretical seismograms of the longitudinal waves. The parameter of the curves is the angle between the boundary normal and the straight line connecting the source with the point of observation; thus, each set of curves defines the directive function of an examined source for the given relationship of constants. The indicated angle is computed from the normal in the medium with the highest velocity. The relationships of the constants are shown on the figures; a, b, and ρ designate velocity of longitudinal waves, velocity of transverse waves, and density, respectively. The form of the primary wave (which such a source would induce in a homogeneous medium) corresponds to the form of the upper curve. This form was chosen for its definition; the computations and the standard curves in [3] permit one to construct similar curves for any other form of primary wave. In the medium of lower velocity there arrives at each point, outside of zone II, a head wave IV and after it a wave III. The distance between arriving waves is assigned tentatively on each seismogram of Figs. 4 and 5.

Figure 5 shows similar seismograms for the case when the source is somewhat displaced into the medium with lower (dashed line) or greater (solid line) velocity (the parameters for this figure are the same as for Fig. 4).

It may be seen in Figs. 4 and 5 that wave III is weaker than the preceding head wave with the positive first arrival. Thus, the distinguishing feature of underground explosions, the positive direction of the first arrival, is preserved even when the explosion is set off near a plane of discontinuity.

LITERATURE CITED

1. A. Lyav, The Mathematical Theory of Elasticity [in Russian] (Gostekhizdat, 1934).
2. G. I. Petrashen', A. S. Alekseev, and B. Ya. Gel'chinskii, "Elementary theory of wave propagation," in the Collection: Questions on the Dynamic Theory of Elastic Waves [in Russian], No. 3, Izd. LGU (1958).
3. G. I. Petrashen' (editor), Data on the Quantitative Study of the Dynamics of Seismic Waves [in Russian] 1-3 (1957-58).
4. V. M. Babich, The Reciprocity Principle in the Theory of Elasticity [in Russian]. In Press.